BASIC DIGITAL PHOTOGRAPHY

BASIC DIGITAL PHOTOGRAPHY

NORMAN BRESLOW,
M.A., B.F.A., B.A.

FOCAL PRESS
Boston London

Focal Press is an imprint of Butterworth–Heinemann.

Recognizing the importance of preserving what has been written, it is the policy of Butterworth-Heinemann to have the books it publishes printed on acid-free paper, and we exert our best efforts to that end.

Library of Congress Cataloging-in-Publication Data
Breslow, Norman
Basic digital photography / Norman Breslow.
p. cm.
Includes bibliographical references and index.
ISBN 0-240-80055-9 (casebound)
1. Photography—Digital techniques. I. Title.
TR147.B74 1990
778—dc20 90-35886
CIP

British Library Cataloguing in Publication Data
Breslow, Norman
Basic digital photography.
1. Photography. Applications in computer systems
I. Title
771

ISBN 0-240-80055-9

Butterworth–Heinemann
80 Montvale Avenue
Stoneham, MA 02180

10 9 8 7 6 5 4 3 2 1

Printed in the United States of America

To my wife, Jill, without whose emotional, editorial, and economic support this book would not have been possible.

CONTENTS

PREFACE

We are living during one of the most exciting times in the history of photography. It is a time of revolution, a time when all the rules of photography are up for review, a time when new skills will have to be learned and some old ones retired as obsolete. This new era arose as a result of the personal computer revolution, and more specifically, from the availability of affordable equipment which allows an image to be put into a computer, manipulated, and then put onto a piece of paper or film. Some comments about traditional photography will help to provide perspective on this new revolutionary age.

Photography can be thought of as a means of accurately representing what exists in nature. Traditional photography does this quite successfully, although it depicts a three-dimensional world in only two dimensions. Photography can also be thought of as a means for self-expression, whereby an individual makes a personal statement about the world in which he lives. Unfortunately, traditional photography does not permit a great degree of this type of self-expression. This may or may not be self evident, and to many readers it may sound like heresy. Certainly there is a diversity of photographic images. Yet if you carefully look at traditional photography, you will find that the greatest changes in imagery result from technologic innovations, and not from unique forms of self-expression. For example, the invention of color film permitted a more natural looking rendition of a scene. Increased shutter speeds allowed motion to be stopped. Small cameras permitted less formal looking images. Fisheye, wide-angle, and telephoto lenses permitted changes in perspective.

While it is undeniable that technologic advances do give photographers some latitude to make unique visual images, the range of personal expression is restricted. Compare the "movements" or "schools" of photography to those found in painting. Where are the impressionist, pointillist, minimalist, cubist, abstract expressionist, hard edge, op, and pop equivalents in photography? Painters have explored these divergent approaches with almost no technologic advances in their field. Painters still use brushes and paints much like those used centuries ago, yet have created numerous new painting styles.

To further illustrate this point, compare the painters of the French Impressionist movement to one another. While the work of Manet, Monet, Renoir, Degas, and Toulouse-Lautrec all share similarities, their work is easily distinguishable. Now look at photographs in the book *One day in the life of America.* Subject matter aside, of the 200 photographers whose work is portrayed, it is often difficult, if not impossible, to distinguish the photographs of one photographer from another.

The question arises as to why there is such a diversity of imagery in painting, and such a lack of diversity of imagery in photography. One reason could be that still photography is not the best medium to allow self-expression. If this is the case, then the reason for this needs to be explored.

Generally speaking, the underlying mechanism of photography is simple: Light reflects off an object and passes through a lens, where it is focused onto a piece of film. The object must exist if it is to be photographed. This is the fundamental difference between photography and painting: the painter can create an image from her imagination. The object may or may not exist, or may exist but not in the form the artist chooses to render it. If a painter wishes to illustrate a dragon, all she need do is close her eyes, imagine one, and then paint her vision. If a photographer wants to photograph a dragon, he must first go out and find one.

Painters are free to express themselves in ways photographers are not. A photographer looks at a scene in terms of the lighting ratio of highlights to shadow, and the effect this ratio will have on the film; whether a telephoto lens should be used to compress depth or a wide angle lens to expand it; what the effect of a large or small aperture will be; whether a fast or slow shutter speed to stop or blur motion should be used. These are some of the creative decisions the photographer must make. Creativity for the photographer depends largely on how well she navigates within the narrow straits of technology that define traditional photography. With the exception of heroic efforts, airbrushing, or having a lithographer make changes to the image, the traditional photographer must assemble all the elements of her image at one time and one place. Once the image is on film, it's finished, with the possible exceptions of making it into a high-contrast print, or posterization, or printing it dark or light.

But what if photography worked on a different system than the one we are used to and take for granted? What if light reflecting off an object passed through a lens but struck something other than film, something magical that would allow the photographer to control his image in ways he had never dreamed of? What if the control of the lighting ratio and exposure considerations were treated as afterthoughts, as minor technicalities that shouldn't be allowed to get in the way of creating the image? What if the photographer were free to concentrate solely on the image, and had almost total control over each element in the scene?

This magical process isn't fantasy, but the promise of computerized photography. When working in computerized photography, the exposed film is only the beginning of the process. All the elements don't have to exist at the same time in the same place. Some elements don't really have to exist at all. Moreover, film is only one possible starting point. An image can be "grabbed" from a TV set, video disk, or TV camera. The photographer may prefer to shoot on video tape: at 30 frames per second, 2 hours of tape yields about 216,000 images to choose from. The tape doesn't have to be developed, but can be viewed immediately, either in the studio or, if out on location, in the camera's viewfinder. Unlike film, the tape can be reused numerous times. Finally, there are cameras that use neither film nor videotape, but small disks which store information in much the same way that information is stored in a computer, and these disks can also be erased and reused numerous times.

The differences between traditional and computerized photography become even more meaningful in the manipulation phase of the process. Regardless of whether the final image is to be used for advertising or commercial purposes, or as fine art, it is at this point that the creativity and unique vision of the photographer gets free rein. The photographer is free to enlarge or reduce all or part of the image; to flip all or part of it left to right or right to left; to flop it top to bottom; to rotate it through 360°; to move or delete objects; change colors; combine the background of one image with the foreground of another; cast shadows in any direction or directions; make objects translucent; stretch or squeeze objects, or stretch one dimension and squeeze the other; make part of a color image black and white; reverse all or some colors; and change perspective/parallax. This is only a partial list of what is possible.

The virtually limitless options and possibilities offered by computerized photography are not theoretical, but tangible. Work on one image does not preclude pursuing another variation of that image. As each new idea emerges it can be integrated with an older one, yet the older can be preserved in its original form. If the new idea does not work, the previous version is still intact.

After the image or images have been computerized, they must be taken out of the computer and put onto film or paper. Getting the image out may involve nothing more than shooting the image off the computer screen, or it may involve high technology. In either event, the photographer ends up with a traditional negative, slide, or transparency. It is also possible to put the image directly onto paper, bypassing the use of film altogether. If the original image was shot on videotape or computer disk, and printed directly onto plain paper, the resulting photographic image would have been produced without the use of film, photographic paper, or photographic chemicals!

Movement from traditional photography to computerized photography is inevitable as time progresses, a point not lost on the giants of traditional photography. Eastman Kodak, Nikon, Agfa, Pentax, Rolliflex, and Canon, to name only a few, have entered this exciting new field, bringing with them a tradition of quality products as well as an understanding of what photographers want and how photographers think.

Although the photographer will probably never be as free to create as the painter, computerized photography provides new worlds to explore, and with a freedom unimaginable only a few years ago. We are currently in a vast area of unexplored territory, and it will be up to people like yourself to begin the investigation into what is possible. The purpose of this book is to help you begin the journey.

Up to this point we have used the term "computerized photography," since it is one that most readers can readily understand. However, for the remainder of this book, we will usually use the term "digital photography," which, as we will see, more accurately describes this new photographic process.

* * * *

The author and publisher of this book used their best efforts in preparing it to see that the information contained in it is accurate. However, a book of this nature depends on information from many sources, and the information may at times be in error. This may come about through unintentional oversights or misstatements, or miscommunications. Additionally, computer graphics is a rapidly changing field, and accurate information may become inaccurate with the release of new products, the removal of products from the market, or because of changes in the specifications of a product. The reader is urged to consult authorized dealers of equipment used in digital photography to ascertain whether that equipment is appropriate for their needs. Neither the author nor publisher accepts responsibility for any damages caused by misstatements made in this book.

ACKNOWLEDGMENTS

A book of this type depends on the help of many people. I would especially like to thank Steve Haiman of Digital Image Company, Westchester, Illinois, a service bureau, for making all the slides used in this book, and for answering many questions concerning scanners and film recorders; Keven Krell and Peter Feit of Computer Support Associates, West Los Angeles, a Truevision reseller, for sharing some of their vast knowledge of computer graphics with me; John Bilotta of Jetgraphix, Los Angeles, and Paul Buxbaum of Barneyscan, Berkeley, California, for scanning a number of the images used in this book; Jim Greer of A-VIDD, Long Beach, California, for the loan of a Canon RC-760 still video camera; Hank Nagashima of Tamron, Port Washington, New York, for the loan of a Fotovix II-X slide transfer device; Laura Malone and Al Merkrebs of Time Arts, Santa Rosa, California, and Heather Louckes of Vision Technologies, Freemont, California, for their help and technical assistance; Tom Brinson of KB Systems in Austin, Texas, the only person to ever end a phone conversation with me by saying "Happy Trails."

BASIC DIGITAL PHOTOGRAPHY

PART I

THE BASICS

CHAPTER 1

A BRIEF INTRODUCTION TO DIGITAL PHOTOGRAPHY

In this chapter I will discuss what digital photography is, and then I will list a number of current applications for digital photography.

DIGITAL PHOTOGRAPHY IN A NUTSHELL

Digital photography involves "digitizing" a photographic print, slide, or other visual such as a painting or drawing. The digitizing process converts or translates the colors and shades of gray in the original image into numbers. These numbers can be manipulated, using a computer, to alter the original image. The image can be made lighter or darker, the colors can be changed, elements of the image can be moved to other locations or removed from the scene, and parts of one computerized image can be combined with parts of another.

Digital photography involves three different steps: The original image must first be changed into numbers, the image is then manipulated, and the image is taken from the computer and put onto film or paper. Each step requires special equipment which will be discussed in this book.

Changing the original image into numbers can be accomplished in a number of different ways. For example, a video camera, videotape recorder, or television set can be connected to the computer by a cable, allowing the pictorial information from these devices to be sent directly into the computer. The computer takes the incoming image and converts it into numbers in 1/30 of a second.

If the desired image is on a photographic print, a scanner can be used. The scanner senses the different colors or grays in the print, translates them into numbers, then transmits the numbers to the computer via a cable which connects the scanner to the computer. Both of these methods result in digitized images.

Once the image has been digitized, the computer can be used to manipulate the numbers, thus changing the appearance of the image. You don't have to be a computer whiz to manipulate the numbers, since computer programs have been developed to make this process very easy, and a lot of fun.

After the image has been manipulated to your satisfaction, you can convert it into a slide or transparency. This is done by connecting the computer to a device called a film recorder, which uses the numbers that make up the computerized image to create a slide. Or you can make a paper print by connecting the computer to a printer.

This process (digitizing an image, manipulating it, and converting the digitized image to a print or slide) can result in a photograph that looks very much like a traditional photograph, water color, charcoal drawing, or literally like nothing yet imagined. The range of possibilities is only just beginning to be explored, and seems to be limited solely by the creativity of the photographer.

USES OF DIGITAL PHOTOGRAPHY

Two features make digital photography appealing: its tremendous flexibility in manipulating images and the speed at which it can generate images. Its speed makes it a valuable tool for communicating ideas within a company and over long distances. Its flexibility makes it ideal for fine art, advertising, commercial and fashion photography—for any purpose for which traditional photography is now used. To help illustrate the incredibly broad range of applications for digital photography, some examples of its uses follow.

Speed is the main reason industrial and in-house corporate photography departments are rapidly switching to digital photography. Many mid-sized and large businesses have ongoing needs for charts, graphs, and other visual aids used for in-house meetings and presentations, sales presentations, and training of new employees. In the past, artists drew the charts and graphs, hand-lettered or set type for tables, and then the industrial photography department would copy the art onto slide film. Many of these companies now use computers to create the visuals which can then be converted to either 35mm slides, overheads, or paper prints. One of the most immediate advantages of this process is speed. In the time it took to make one chart by hand, a computer can make many. It is now even possible to use video projectors to project the computer image directly onto a large screen, like a slide is projected. This new technique saves not only time, but also the cost of film and processing.

Other industrial photography tasks, such as documentation of manufacturing processes, are also benefiting from the increased speed characteristic of digital photography. For example, suspected malfunctions in machinery can be videotaped. The video image can be put into a computer, where the image can be enhanced or enlarged, arrows or other marks can be added to draw attention to important visual information, and then the image can be sent to a paper printer which makes a dry color print. The entire process of computerizing the image and making a print can take as little as 10 minutes.

Given that digital photographs are essentially a sequence of numbers, they can be transmitted quickly from one computer to another over long distances via telephone lines. Digital photography allows information to be sent to corporate headquarters from distant locations for rapid decisions. For example, if difficulties necessitating corporate intervention arise at a building site, a photographer can take pictures with a digital camera and send the image, via telephone lines or cellular technology, directly to the corporate board room. Many department stores now use digital photography to transmit pictures of products to corporate headquarters for immediate approval of purchases, which again saves company time.

News photography also benefits from the speed of digital photography. Fast-breaking news stories can be shot with digital photography equipment and the images transmitted directly to the newsroom where prints can be made by electronic printers in a matter of a few minutes. The time required to transport the film to a photo lab for processing and printing is saved, which can make the difference between having visuals for an important story by deadline or not. For example, during the student uprising in Ti'anamen Square, satellite transmission from China was cut off. However, the telephone lines were left functioning. Some U.S. television networks converted videotape images into still digital photographs which were transmitted over the telephone lines to TV stations in the United States, where they were broadcast as standard video signals. All of this happened in a matter of minutes.

A truly extraordinary feature of digital photography is that it enables discussion and modification of visual images over long distances. For example, it is possible for a photographer to transmit a digital photograph over telephone lines to a computer in an advertising agency in another city. The photographer can actually manipulate the image appearing on the advertising agency's computer screen, showing what the picture would look like darker, in pastels, or in a different composition, while talking to the art director on a telephone. The art director could ask "what if" questions and the photographer could make instant visual adjustments. This ability not only helps the creative process, but also eliminates misunderstandings which sometimes arise when talking about visual images without actually seeing them.

Besides speed, digital photography offers virtually limitless flexibility in creating imagery. Many visuals, especially those used in advertising, commercial and fashion photography, are meant to grab the viewers' attention. It is up to the photographer to make the product stand out. The digital photographer can create images that would be difficult or impossible to shoot with traditional photographic techniques. Moreover, many alternatives can be explored in brief periods of time: A full color image can be changed into a pastel image with the touch of a few keys on the computer keyboard; the resulting image can in turn be modified, by increasing or decreasing the contrast, lightening or darkening the image, or changing it into a charcoal-like drawing.

The digital photographer can create many variations on a theme, and in doing so, has more precise control over the photographic process. In traditional photography, techniques such as posterizations, solarizations, or high-contrast effects are often hit or miss manipulations, and each attempt could take quite some time. In digital photography, a posterization may be accomplished in seconds, and if it is not appropriate for the job at hand, it can easily be modified or abandoned, as the time investment is minimal.

Digital photography can be applied in all areas of traditional photography, allowing a previously undreamed of control of the image, regardless of the subject: wedding, landscape, architectural, food, fashion, sports, etc. Every portion of the digital image can be individually manipulated, making digital photography the ultimate retouching medium. If a model's eyebrow needs to be touched up, the digital photographer can do it in seconds. If the light coming through a window in an interior shot is too bright, it can be darkened, while the remainder of the image is left untouched.

In addition to its use as a retouching tool, digital photography is an extremely powerful special effects generator. Its ability to merge portions of one image with another, for example, merging the sky from one slide with the buildings from another, adds to the fine or commercial artist's ability to create unique visualizations.

Digital photography can also be applied to do things totally beyond the domain of traditional photography. For example, computerized architectural renderings of proposed buildings can be inserted into photographs of the actual building site, to make more meaningful presentations to clients or prospective financial backers. Landscape architects and gardeners can merge photographs of clients' houses with photographs or paintings of shrubbery, lawns, and flower beds to present their ideas. If the client wants to see what his house would look like with different shrubs, the computerized image can be changed as the client watches. A color printout can be made in a matter of minutes for the client to take home to study.

Computerized photography is being used by plastic and reconstructive surgeons to show the client the end result of the proposed surgery. Cosmetics companies can show the aging effects of the sun, by photographing customers and then using a computer program to show what the individual will look like in 20 years, based on answers to questions about the customer's skin type, hours of exposure to the sun, etc.

Digital photography is being used to catalog and allow for viewing, via computer-televised imagery, extremely rare and fragile works of art, which otherwise would not be displayed. For example, the Getty Museum in Los Angeles is using digital photography to copy its collection of illuminated manuscripts onto videodisks/videotape for viewing.

These are but a few of the uses of digital photography. The remainder of this book introduces you to the equipment and techniques that are used for making these new types of images. Once you have the tools and a basic familiarity with the techniques, you will find that you are limited in your photographic imagery only by your imagination.

CHAPTER 2

ESSENTIAL VOCABULARY AND CONCEPTS

This chapter is written specifically for those readers who have no, or only limited, experience with computers. It introduces you to the basic vocabulary of computing. If you are familiar with computing, you may wish to skip to Chapter 3. If you are not computer literate, this chapter will not make you into a computer whiz, but it will be your first step in that direction. The vocabulary and computing concepts I introduce here will be used in the following chapters to explain what equipment is available for making digital photographs, what it will enable you to do, how it is used, and what choices you will have to make when choosing equipment. The definitions and examples have been selected to be as straightforward and painless as possible, and some may tend toward oversimplification. Too much detail, too soon, can be overwhelming. I suggest that you do not try to memorize the information in this chapter. Instead, reread it from time to time, and refer back to it when necessary.

The vocabulary in this chapter is not given in alphabetical order; rather, earlier words and concepts are used to help explain later ones. Use the glossary in the back of this book for a quick reminder of the meanings of terms.

Above all, keep this in mind: THERE ARE NO TESTS. Read this chapter for the information it contains, and come back to it from time to time to remind yourself of particular concepts. The purpose here is not to teach you how to actually use a computer, but to give you an idea of what a computer is, how it works, and what you can do with it.

This chapter is divided into four sections: First we will discuss the usefulness of jargon in general; second, I will offer some strategies you can use to learn more computer jargon in your journey toward computer literacy; third, we will begin our discussion of those computer terms that will be used throughout this book; and finally, there will be a brief summary of the information covered in this chapter.

JARGON

Regardless of what we do in life, we all learn one or more specialized languages, words that are meaningless to others who do not share our interests. Lawyers talk about briefs and writs, typographers of ems, picas, and fonts, and photographers talk about watt seconds, bellows extension factor, and f/stops. Such jargon develops as a shorthand way of transmitting meaningful information quickly and accurately. Consider the following photographic instructions:

1. In order to best use the piece of glass that makes everything look very small on that big camera, which makes everything look upside-down and that uses single pieces of film which you put into it one piece at a time, you must remove the thing that looks like the central part of an accordion which is probably made out of semi-flexible vinyl or leather, and replace it with a different thing that looks like a bag and has openings at opposite ends, which is probably made out of very flexible vinyl or leather.

2. To make the best use of a wide-angle lens on a view camera, replace the standard bellows with a wide-angle bellows.

While both sentences attempt to convey the same information, the second one does it faster, more quickly and more accurately, and only requires the knowledge of three specialized concepts: Wide angle, view camera, and bellows. With this basic knowledge, the concept that a wide angle bellows should be used with a wide angle lens on a view camera can be quickly communicated. Given the brevity and power of specialized words, it's no wonder that specialized jargons develop. Without this vocabulary, things get messy.

Those involved with computers also have developed a jargon of their own. Because computers are relatively new, this jargon is still in a phase of rapid development. New words and phrases seem to be coined daily. Some computer jargon is obscure, and will eventually lose popularity and become archaic. One word ripe for extinction is *boot,* which simply means to turn on a computer. The word *boot* was originally used by early computer users because they recognized that when a computer is first turned on, it is in a state of amnesia. It must find information to relearn how to perform certain functions. This situation reminded the early computer users of the saying "To pull yourself up by your own bootstraps," and so the term *boot* came into use. Although boot has deep meaning for some, it is an example of the misuse of jargon, since it conveys little practical information, and does a lot to confuse novice computer users. While the term is falling out of use, being replaced with the more meaningful "turn on your computer," it can still be found in computer manuals and books.

Other computer jargon is more straightforward. For example, *backup* simply means to make a copy of something that is important in order to protect yourself if you should accidentally lose or damage it. This concept, to *backup* important information, assumes certain knowledge on the part of the computer user. Just how does one make a backup? Is there more than one way? What exactly do I do? A photographic analogy would be the experienced photographer telling a novice to develop film. A simple concept, but it assumes knowledge of (1) what film hangers or reels are and how to use them; (2) time and temperature control; (3) dilution of developer, stop bath, and fixer; (4) a final wash; (5) what each chemical does; (6) when to turn the lights off and when it's all right to turn them back on; (7) what agitation marks are and how to avoid them; (8) how to remove excess water from washed film to avoid water marks, etc. The understanding of jargon often depends on understanding other jargon.

LEARNING JARGON

In order for you to effectively use a computer to make photographs, it should be apparent that a specialized vocabulary or jargon will have to be learned. The reader has a number of options for doing this.

1. Extension courses about computers can be taken at colleges and high schools, and courses are given by computer stores and adult education centers. Formal courses such as these may run the gamut from "Learning All about Personal Computers" to "Advanced Programming for Computer Professionals." Such courses are valuable, although more so for those who appreciate formal or structured learning environments; those who prefer to jump right in and get their hands dirty may find such courses frustrating. Most courses beyond the "Introduction to Personal Computers" level are geared to specialized interests. Most courses in computer graphics are geared either to computer programmers, or to commercial artists who want to learn computer drawing and design skills, and not to photographers who want to learn computer photographic skills. Although computer classes designed for commercial artists can still be valuable to photographers, especially since such courses will cover many areas directly related to digital photography, much time and effort will be dedicated to tasks and skills of little interest to most photographers. The frustration level of learning how to set type with a computer may be too much for the individual who has no particular interest in that area.

The advantage of taking courses is that an expert (the teacher) is available to answer questions. Beginning computer courses can be very helpful, especially after having gained some knowledge from a book such as this, if for no other reason than that they put you in contact with an expert who may be able to clarify some of your more specific questions.

2. A second approach to learning about computers is to locate books on the particular subjects of interest. At present, no other "how to" book on digital photography is available, although this will change as public interest in this topic increases. Just as there are problems associated with formal classes, there are problems when learning about computer applications through a book.

First, you must locate a book that is written on a level appropriate for you. Having found your book, you will find that a great deal of effort is involved in absorbing the information. Often problems arise due to ambiguities or the necessary incompleteness of the information.

An article by John C. Dvorak, "Crazy Mistakes, Part 1" in *PC* magazine (March 28, 1989) attests to problems which can arise when one is first learning about computers from books or manuals. The article contained the following anecdote:

A man called the store he bought his computer from because he could not get it to work. He followed all the instructions in the manual to the letter: "Put your disk into the disk drive and close the door." (We will discuss disks and disk drives in detail later in this chapter.) The man inserted the disk into his disk drive and then got up and closed his office door. Why, he wanted to know, wouldn't his computer do anything? Apparently the person who wrote the manual had a lapse in clarity, and did not adequately explain what "door" meant.

Finally, any computer book, including this one, can only give you information, but you still must get a computer to begin to put that information into practice.

3. The third way to learn the jargon of computers and computer graphics is to jump into it. Buy equipment, read the manuals, call up the "customer support line" most manufactures provide with their products, call the person who sold you the equipment, buy some books that cover the areas you are having difficulty with, and keep at it. Make friends with someone who knows more about computers than you do. Pick her brain. This is how I learned what I know about computers and computer graphics. Many books. Many telephone calls. Many, many magazine articles. And many frustrating hours spent in front of my computer trying to figure out why it wasn't doing what I told it to do. This method is not recommended to those who do not take a certain perverse joy in being frustrated, nor for those who don't feel exhilarated when they solve a mystery.

It is doubtful that any one of these three approaches, by itself, is the best way to learn digital photography, especially if you do not have a background in computers to start with. A combination of these methods will probably work best. If I have made it appear that becoming "computer literate" will take some investment of time and effort, I meant to. But you will find that the task is not insurmountable, and by using some of the methods described you too will be among the computer literate sooner than you think.

COMPUTER JARGON AND CONCEPTS

A Word about Anthropomorphising

It is impossible to talk or write about computers at any length without attributing to them certain human characteristics. It is common to say: "When you type in this command, the computer *thinks* that...", "The computer *wants* to ...", or "The computer *knows* that... ." Computers do not think, nor do they want, or know. They follow instructions. They do repetitive, boring tasks rapidly. They are not human, although we seem to think of them in human terms.

Background

A PC (for *P*ersonal *C*omputer) is a computer designed to be used by one person at a time (Figure 2.1). This differentiates it from other types of computers (e.g., mainframes, minicomputers) which have a central computing unit that a number of different people can use at the same time. This book will concentrate on PCs, and unless specifically stated, assume that all references to computers and computer accessories are for personal computers.

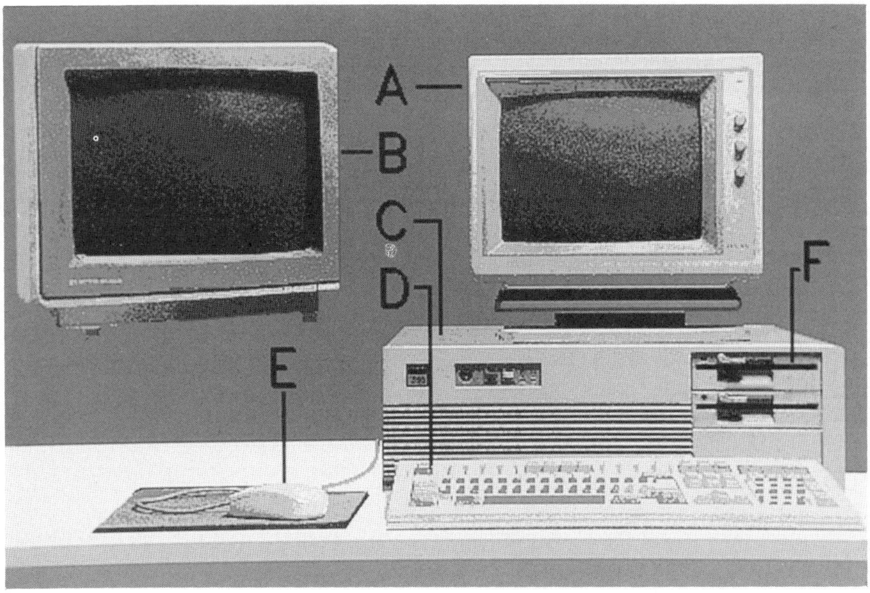

FIGURE 2.1 A computer and equipment: (A) standard computer monitor; (B) digital photography graphics monitor; (C) computer; (D) keyboard; (E) mouse; (F) floppy disk drive. Image made with TIPS and utility program.

Architecture is a computer term you will rarely hear or read, and it is a term with more than one meaning. In its first meaning, architecture describes how a computer is physically designed, and so can be thought of as engineering specifications. The second meaning is more important to this book. It describes the philosophy of the designers of the computer, and usually refers to the choice of closed architecture versus open architecture.

One philosophical approach is to design a computer as a self-contained unit. This type of computer is referred to as having *closed architecture*. The manufacturer decides what the computer should be capable of doing, what features it should have, what component parts it will have, and builds all of this into the computer. The manufacturer has built, for all practical purposes, a sealed box. If market research shows that a need exists for a different type of computer, or if technological breakthroughs are made which will improve the computer, then a new model can be brought out. The Apple computers and the first Macintosh are examples of closed architecture computers.

There are both advantages and disadvantages to the closed architecture approach. Closed architecture computers are usually easy to learn to use, because the manufacturer can construct ease of use into the computer's overall design. A second advantage is that most, if not all, of the sometimes bewildering choices the first time buyer is faced with have been made by the manufacturer. A third advantage is that there are no *compatibility* problems. Everything in the computer works with everything else, because it is built as one integrated unit.

These advantages must be weighed against the disadvantages. First, there is usually a rapid obsolescence factor. If new technology makes a better computer possible, the older model is discontinued and replaced with a newer one, because there is usually no way to *upgrade* the older model to have the same features as the improved one. It can't grow. It will always be what it was designed to be when you bought it. A second disadvantage is that the computer is only capable of doing what the manufacturer decides it should do. If the manufacturer does not feel that digital photography is important, then that capability will not be built into the computer, even if the technology exists. The trade-off when buying a closed system is one of initial ease of use versus long-term usefulness. You can think of a closed architecture computer as a camera that has a fixed, nonremovable lens. It might be a good lens, but you do not have the option of replacing it with a different one.

The alternative to closed architecture is *open architecture*. The manufacturer literally gives access to the "innards" of the computer to anyone wishing to design or build parts to work with it. This is done by leaving room inside the computer where additional equipment can be placed. This approach takes the saying "two heads are better than one" to an extreme. Instead of one company deciding what the computer buying public wants, or what is feasible, a potentially unlimited number of entrepreneurs are free to "improve" the existing computer, or find a market and design specialized equipment for it. If one company decides that a computer market exists for digital photography, they can design equipment to work with an open architecture computer, and then sell the equipment to those who own that computer. IBM PCs, XTs, ATs, and the Macintosh II, and to a lesser extent, IBM PS/2s are examples of open architecture computers.

As with the closed architecture system, there are both advantages and disadvantages to an open architecture system. One of the advantages of this approach is that obsolescence can be delayed. As innovations are introduced to computing, one or more companies will find a way to upgrade existing open architecture computers to the new standard. The cost of upgrading is usually significantly less than purchasing a new computer, and you have the option of deciding which new features are worth an upgrade and which are not. Another advantage is that you have much more control over the features of your system. Because so many *add ons* are available, you are free to pick and choose among them. Additionally, it is possible to have specialized equipment built to your specifications, although this may get expensive. Another advantage is that there is competition among *third party* manufacturers, which means that prices are kept low.

The flexibility that makes open architecture computers so desirable also causes problems. In putting together a system, you will be faced with seemingly endless options to choose from. These options can be confusing. Once equipment has been selected, you will have to learn how to use several different components. These components may have been designed by different manufacturers, each following different, and possibly conflicting, logic. Finally, not all components work well together, so compatibility becomes an issue; a new piece of equipment may fail to work in the presence of other existing equipment, or it may cause the older equipment to malfunction.

A *clone* is a computer, computer attachment, or computer program that is designed to work like an existing computer, computer attachment, or computer program. (More about programs later.) Clones are also called *compatibles, work-a-likes,* or *look-a-likes.* Clones of closed architecture computers are rare, since they usually would have to violate numerous patents, while clones of open architecture computers are numerous since there are usually less patent restrictions to work around.

Memory

One of the more important computer concepts is that of the byte (pronounced "bite"). Most computer users think of a byte as any character which can be typed into a computer. By character, I mean any letter of the alphabet, any digit from 0 through 9, any punctuation mark, a space, or special symbols such as > or \. For our present purposes, this understanding is sufficient.

A byte is not only a character, but a unit of measurement. A word with five letters in it contains five bytes. A book of this length may contain 750,000 bytes. Of all the computer concepts found in this book, you will probably find yourself using the concept of the byte, as a unit of measurement, the most often.

A kilobyte, usually abbreviated K or KB, is 1024 bytes, but for many purposes you can think of a kilobyte in its rounded form of 1000 bytes. A megabyte, which is usually abbreviated MB, is one million bytes (rounded off), and a gigabyte, or GB, is rounded off to one billion bytes.

Just as a word is made up of letters, a byte is made up of bits. While words have a variable number of letters, all bytes (as far as this book is concerned) have eight bits. You can visualize a bit as an on-off light switch. In one position, the light is switched off, in another position it is switched on. These are the only two possible positions. A bit represents a positive or negative electric charge. It can be either negative (off) or positive (on).

Imagine eight light switches in a row. Some are in the off position, and some are in the on position. Now substitute a sequence of eight electric charges, some negative and some positive. The computer interprets these off and on electric states as either the digits 0 (negative charge) or 1 (positive charge). These are the only two digits the computer can understand. Because of this, computers work on a binary number system, instead of the decimal system we are used to. The word *bit* stands for *bi*nary digi*t*. Below is a visual example:

Eight light switches:	on	off	on	on	on	on	off	off
Eight electric charges:	+	-	+	+	+	+	-	-
Eight bits:	1	0	1	1	1	1	0	0

The computer interprets the negative and positive electric charges as binary digits (bits), and groups eight bits together. The sequence of the eight 1s and 0s identify one *byte* from another. For example, one sequence may represent the letter J, while another may represent the letter j. There are 256 different possible 0-1 combinations the eight bits can make ($2^8 = 256$), and so a computer can identify 256 different characters. This is a sufficient number to represent all of the UPPERCASE and lowercase letters of the alphabet, the digits 0–9, all the punctuation marks, a symbol used by the computer for a space, special characters such as * and |, other symbols used specifically by the computer, and still leave plenty of possible symbols for future uses.

Chips

A computer chip looks like a small domino (without the dots) which has a number of staple-like legs protruding from its sides (Figure 2.2). Inside the chip is an incredibly complex series of etched pathways, which carry and store electric current. Different chips are designed to perform different functions. For example, some chips are specially designed for use in graphics, and others for use in communicating over telephone lines. Some equipment allows for increased productivity by allowing the owner to add additional chips after the equipment has been purchased. The type of chips that the owner of such equipment may wish to *install* are specified by the manufacturer, and so the choice is whether to add more chips or not, but usually not what type of chips. This is similar to buying an optional battery pack for a camera. You can decide whether or not to buy the option, but if you do, you are limited to the type of batteries the pack can use.

If a computer can be said to have a brain, the CPU (for *C*entral *P*rocessing *U*nit) would be it. The CPU is a special chip, found on the motherboard of the computer, which takes the commands given to the computer (either from the keyboard or from a computer program), interprets them, and then executes them. It is a central clearing house, keeping tabs on what the different parts of the computer are doing, and instructing them what to do next.

RAM (for *R*andom *A*ccess *M*emory) is the computer chip in which most of the work done by a computer takes place. Although the computer has only one CPU, it has many RAM chips. The more RAM chips the computer has, the more

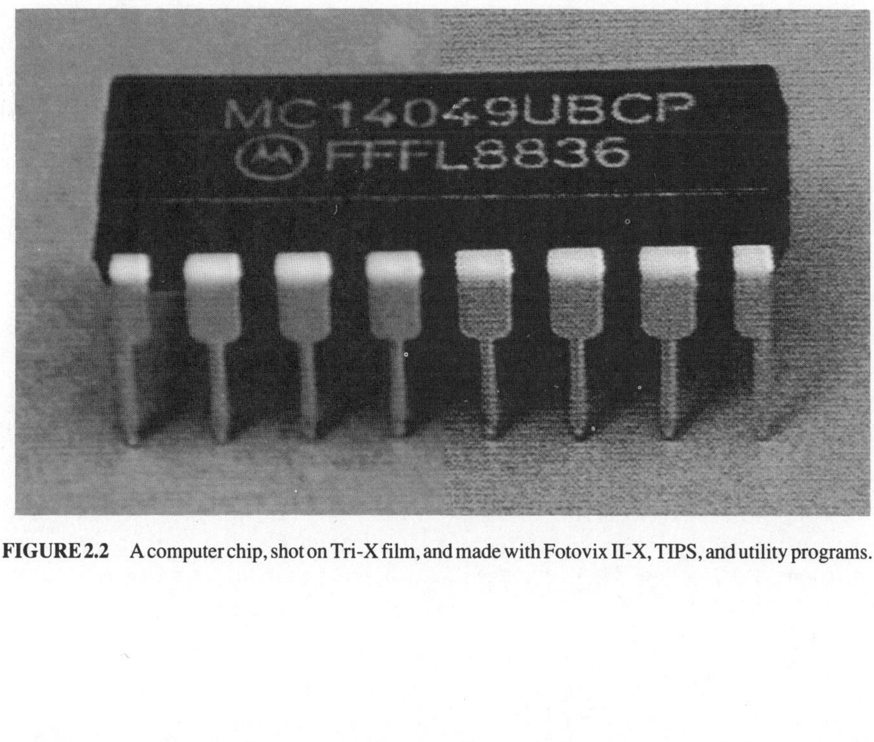

FIGURE 2.2 A computer chip, shot on Tri-X film, and made with Fotovix II-X, TIPS, and utility programs.

information the computer can manipulate. While there are a number of different types of RAM chips, all share some things in common.

1. All RAM chips are capable of both (a) receiving, manipulating, and storing information such as keystrokes typed into the computer, and (b) sending that information to the computer screen, printer, or some other device. When information is put into RAM, you are *writing* to it. When information is sent from RAM, you are *reading* from it.

2. The information in RAM is not permanent. When you shut off your computer (or if your cat pulls the plug from the wall), all the information in RAM is lost. The reason the information in RAM is *volatile* is because it consists of only positive and negative electric charges. (Remember *Bit?*)

IBM and compatible computers have a maximum amount of RAM which can ordinarily be used to process information. This amount is 640KB. Think of this as meaning that the computer can manipulate up to 640,000 characters at a time. When personal computers were in their infancy, this was considered a huge amount of RAM. Many of the first personal computers that were sold in the early 1980s had 16KB or 32KB of RAM which could be used to manipulate information, and so 640KB seemed like more than enough. But times have changed rapidly. Today, 640KB of RAM is considered barely adequate, but the original design of the IBM and IBM-compatible computers limits the amount of RAM the computer should be able to use to 640KB. We will refer to this RAM as *standard* RAM, because some methods have been designed to fool the computer into using more RAM than it was designed to use.

Storage

If most of the work done by the computer is done in RAM, and if RAM is not permanent but disappears when the computer is shut off, then it would be self defeating to do anything with a computer that couldn't be finished in one sitting. One solution might be to leave the computer on at all times, and hope there is never a power outage. A better solution is to have some type of storage device that is permanent, and that can take the information in RAM and save it for later use. A number of such devices exist, with the most common being the floppy disk, (also called floppy diskette, floppy, disk, or diskette).

Figure 2.3 shows a floppy disk. While they come in a number of different physical sizes, the two most commonly used are the 5.25 inch and the 3.5 inch disks. The 5.25 inch disk (pronounced "five and a quarter") consists of a flat, thin, relatively stiff, cardboard-like outer container or *jacket* which has a number of openings in it. One, which is oval-shaped, permits information to be either written to (put onto), or read from (taken from), the floppy disk. The 3.5 inch disk (pronounced "three and a half") differs in that it is smaller, thicker, and its outer jacket is made of stiff plastic.

If the 5.25 inch disk is only relatively flexible, and the 3.5 inch disk is stiff, why are they called "floppy"? Inside the jacket is a thin layer of magnetic media, much like videotape. If you imagine a 5 inch diameter circle the thickness and

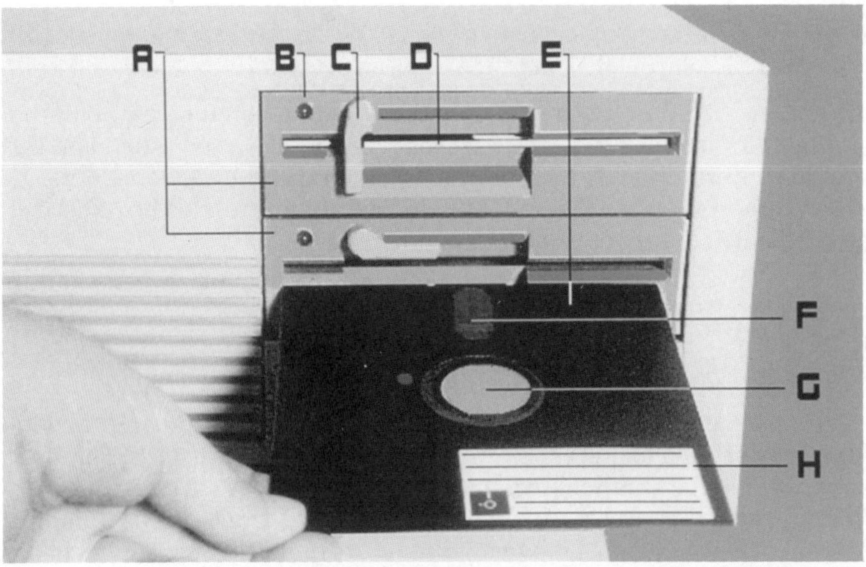

FIGURE 2.3 A floppy disk and a floppy disk drive: (A) two floppy drives; (B) light that indicates that drive is in use; (C) latch that locks floppy disk into position; (D) disk locked inside floppy drive; (E) floppy disk; (F) oval hole in disk jacket exposing magnetic media; (G) hole in jacket used by floppy disk disk drive to spin inner magnetic media; (H) label used to identify contents of disk. Image made with TIPS and QFX utility program.

consistency of videotape, you will be able to understand why the name "floppy" applies. It is the outer jacket that helps the magnetic media hold a flat shape.

The floppy disk works in much the same way videotapes work, in that information is stored and retrieved in magnetic form. And, as with videotape, the information stored on the disk can be *erased* or *overwritten*. This is handy if you no longer have use for the information on the disk and want to use it for new information, but can be disastrous if the information is accidentally destroyed.

Disks can also be compared to records. Disks are divided into circular tracks, which serve the same function as the grooves in a record. But here the similarity ends. Grooves in a record are in a spiral shape, starting from the outer-most edge and ending somewhere near the middle. Tracks on disks are circular, like concentric rings in a tree trunk. Another difference is that you can see and feel the grooves on a record, but on a disk the tracks are magnetic, and so are invisible to the eye or touch. Besides tracks, disks are also divided into magnetic triangular or pie-wedge sectors. These sectors aid the computer in rapidly locating information on the disk.

Different types of disks can store different amounts of information. Some 5.25 inch disks can store a maximum of 360KB of information, that is, 360,000 bytes (or characters), while others can store 1.2MB of information (1,200,000 bytes). Some 3.5 inch disks can store 700KB of information, while others can store 1.44MB. A photographic analogy would be different lengths of 35mm film. Some rolls will allow you to shoot 12 pictures, others 24, and still others 36. But unlike a 35mm camera which can accept any standard length of 35mm film, different computer systems may require the use of a particular type of disk. Moreover, disks of the same physical size (e.g., all 5.25 inch disks) that have different storage capacities usually need their own specialized equipment to work properly. If your computer is set up to use only a 5.25 inch disk that has a storage capacity of 360KB, you will not be able to use a 5.25 inch disk with a storage capacity of 1.2MB.

Disks are usually sold in boxes of 10, although it is possible to buy them in bulk, usually in quantities of 25, 50, or 100. The prices of disks vary, with low storage capacity 360KB disks costing less than high storage capacity 1.2MB or 1.44MB disks. Prices also vary within the same "family" of disks. Off-brand 360KB disks can cost as little as $6 for a box of 10, while name brand 360KB disks may cost as much as $20 a box. Off-brand high-capacity 1.2MB disks may cost $12 for a box of 10, or $40 for a brand name box. (I, like many other computer users, buy the least expensive disks I can find.)

Floppy disks are used not only to make information in RAM permanent, but also to allow information from one computer to be transported to another. If you have one computer at home and another at your office, you can *transport* information from one place to another by carrying it on a floppy disk, or you can mail information contained on a disk to a person who uses the same type of computer. It also allows information to be stored in a different place for safe keeping.

Although disks serve many essential functions, they do have some disadvantages. First, they have limited storage capacity. Although 1.2MB represents over 1,200,000 characters, enough to hold easily a book of this length, computerized photographs can get quite large (as measured in bytes). Depending on the system you choose to use, your images may range from 400KB to 50MB or more. Another problem with disks is that they work more slowly than some other storage devices. Speed in computing is a relative term. A floppy disk may take twenty seconds to perform a task while another device can do it in three seconds. Seventeen seconds may not sound like a long time, and it's not (especially if it would take you a few hours to do it by hand). But if your computer application uses the disk often to perform its task, the seventeen seconds add up.

A disk drive or floppy disk drive is to a disk what a videotape recorder is to videotape. A disk drive contains the magnetic read/write head or heads that put information onto the disk and get information from it. It also contains, or is attached to, a disk drive controller, which electronically guides the read/write heads to different locations on the disk. Floppy disks are random access devices, and information from any part of the disk can be read without having to read unwanted information. This is similar to picking up the arm of a record player and moving the stylus to a different part of the record, thus avoiding listening to unwanted music. Another reason a disk drive needs some sort of electronic guidance device is that the tracks on the disk are circular. Without such a device, the read/write heads might be doomed to continuously circle one track forever.

To use a floppy disk, you slide it into the disk drive, with the oval opening entering first (Figure 2.3). On the outside of the floppy disk drive there is usually a latch or other mechanical device that will lock the disk into position. Sometimes this device is a "door-like" closure. When the floppy disk is locked into position, the floppy drive clasps onto the magnetic media through the central hole in the jacket. The floppy drive spins the media (the outer jacket does not spin, only the inner media), and when the computer is given appropriate instructions, the read/write heads in the floppy drive, maneuvered by the disk drive controller, seek out the appropriate position on the floppy disk through the oval opening in the disk's outer jacket. Depending on the instructions given to the computer, information is then either written to the disk, or read from it.

All personal computers have at least one disk drive, if for no other reason than to allow the use of store-bought programs. If you buy a game at a computer store, it comes on a floppy disk, and you need some way to get the game into the computer, to play it. The floppy disk drive is the entry point.

While a floppy disk is needed to put store-bought programs into your computer, and is helpful in storing information from RAM, making backup copies of important information, and transporting information from one computer to another, it suffers from relatively low storage capacity and relatively slow working speed. A different type of disk is available that overcomes these shortcomings. This is the hard disk (also called *hard drive* or *fixed disk*). Hard, because the magnetic media is not on an ultra-thin base but on a rigid metal or "hard" backing. Fixed, because the disk is permanently attached to the disk drive. You do not insert and remove the disk as you do with floppy disks. Instead, the disk is sealed inside a protective box, which is usually located inside the computer cabinet.

Hard disks work very fast. They spin at 3000 revolutions per minute (compared to the floppy's 300 or 360 rpm), and the read/write magnetic heads move very rapidly across the surface of the disk. Hard disks can accomplish this rapid movement of the heads because, unlike with floppy disks, the hard disk heads do not come into contact with the magnetic media; rather, they "float" a tiny fraction of an inch above it, and so there is less friction to overcome. (The read/write heads of a floppy drive, on the other hand, actually make contact with the disk.)

Most hard disks used at this time are 20MB, 30MB, or 40MB in storage size, although 65MB, 80MB, and 110MB disks are becoming readily available. One reason hard disk drives can store so much information is that they can contain more than one disk. A single disk may store 5MB or more on one of its sides and the same amount on the opposite side. A number of disks can be stacked on top of each other inside one hard disk drive, with only enough room between them to permit the relatively flat read/write heads to operate on each surface.

While hard disks overcome the problems of the slow speed and low storage capacity of floppies, they suffer from two problems. First, they will eventually get filled to capacity. At this point you will either have to transfer some less often used information to floppy disks, thus freeing space for newer, more important information on the hard disk, or you will have to buy another hard disk. At $250 to $2000 per hard disk, (depending on such factors as the type of computer it is used in, the number of bytes the drive can store, called *maximum storage capacity*, and the speed that information can be found on the disk, called *access time*), this can get expensive.

A second problem with hard disks is that they do not enable easy transfer of information from one computer to another. Since you are not meant to insert or remove the hard disk, it is usually installed inside your computer, and is difficult to get to. It is not the kind of thing you just pull out of the computer. Since they weigh a few pounds, and are both bulky and delicate, mailing them to someone is not recommended. If you want to send a photograph that is on a hard disk to a friend in another city, you could transfer the image from your hard disk to a floppy disk and send it through the mail. If your photographic image is only 500KB, this is not a problem, since it will easily fit on a 1.2MB floppy disk. If your image is 2MB in size, you can break up the image into two sections, and put part of the image on one floppy disk and the remainder on another. But if you are working on a very involved image, which may be 20MB or greater in size, breaking the image into 15 or more parts can be time consuming and a bit frustrating. Your early pictures won't entail such large files, so this will not be a problem at first. However, as your pictures become more complex, this issue will become a concern.

Hard drives are sometimes referred to as *mass storage devices*, since they are capable of storing much more information than floppy disks. Some mass storage devices are designed to overcome the problem inherent in the permanent mounting of the fixed disk in the disk drive; such removable mass storage or removable hard disks can be removed and inserted in much the same way as floppy disks. They can be used to transport large amounts of information, and to provide a virtually limitless storage capacity. They usually have capacities ranging from 5MB to 20MB; if you had such a disk and it were to become full, just remove it from the drive and insert another one in the same way as you would insert a floppy disk. Such devices usually cost between $1,000 and $2,000, and additional disks usually cost between $50 and $150, depending on the type of drive and the disks' storage capacity.

A WORM (*W*rite *O*nce *R*ead *M*any) drive is a special type of drive, which utilizes disks similar to CDs (compact disks) used with compact disk players. As a matter of fact, it is possible to get WORM drives that can also play CDs, assuming of course that your computer is attached to a stereo system. While you can put information onto a WORM disk, you cannot erase it, because the information is literally etched into the disk by a laser beam. It is truly permanent, assuming that you do not physically damage the disk. This device is not practical in most circumstances because once the disk is full you cannot overwrite it (write new information over old information) with more valuable information. The advantage of a WORM device is that it has a large storage capacity, usually 500MB or greater. This, and the permanence of the stored information, make it an ideal choice for *archival* storage. A large company can keep all of its payroll data for a year on a single WORM disk, and a photographer can keep as many as 1000 photographs on one. At present, such devices are costly (usually about $3000), but as with other devices that are expensive when first introduced to the computer market, prices will drop as more manufacturers enter the field. Additionally, currently under development are devices similar to WORMs, which allow information to be erased or overwritten. These are usually called *optical drives.*

Movement Among Disks

Your computer is not limited to only one type of disk drive. Let's say you have a computer that has a 360KB floppy disk drive, a 1.2MB floppy disk drive, and a 20MB hard disk. You may want to take information from your 360KB drive and put it into RAM so you can work with it, or take information from your fixed disk and store it on a 1.2MB floppy disk. How do you tell the computer which drive you want to read from, or which you want to write to? This is easily accomplished by assigning each drive a different letter of the alphabet. Older computers had a limited number of possible drive letters or drive specifiers that could be used, but most computers made today will allow up to 26 different drives to be used. In practice though, most people use between two and four drives.

By convention, floppy drives are given the letters A and B. If you have only one floppy drive, it would be called "drive A," and if you have two, one would be called "drive A" and the other "drive B." If your computer has one fixed drive, it is labeled "drive C" by convention. If you have a second hard drive, you would call it "drive D." If you also have a removable mass storage drive, you would call it "drive E." When you want to do work on a floppy disk, you would tell your computer to turn its attention to drive A (or drive B), and if you later wanted to work from your fixed disk, you would tell your computer to turn its attention to drive C or D.

When using a computer, you are always working with a drive. The drive you are currently on at any given time is called the *logged drive*. While using your computer, you may want to change from one drive to another. You may be logged on hard drive C, but you may want to use information you have stored on a floppy disk. After putting the floppy disk containing the information into your floppy disk drive, you would tell the computer to change from drive C to drive A (or drive B, or whatever drive specifier you have assigned to that particular floppy drive). Before you switch drives, you are *logged on drive C*; after you switch, you are *logged on drive A*.

To switch drives on IBMs and IBM compatibles, you type the letter (drive specifier) of the drive you want to *access*, or log onto, followed by a colon (:), followed by a carriage return. To change from drive A to drive C, you would type C: and press the carriage return key on your computer keyboard. The computer will do the rest.

It is important to understand that whenever a personal computer is on, it is working with a disk drive. When the computer is first turned on, it must locate information it needs to begin to operate. This information may be stored on a hard drive, or on a floppy disk. In either case, the computer searches for a drive that contains the needed information. Once the information is found, the computer is capable of following instructions. If you do not tell the computer to switch to a different drive, it will stay logged onto the drive where it found the needed information.

If you have more than one disk drive, it's not difficult to forget which drive you are currently using. This could cause annoying problems. The computer helps remind you by printing the letter (drive specifier) of the currently logged disk (the disk you are currently using) on the left side of the computer's screen, usually, but not always, at the top. On IBM computers and IBM clones, the drive specifier will be followed by a > symbol, which is called a *prompt*. If you were on drive C, you would see C> (pronounced "C prompt"). This tells you two things. First, that you are logged on drive C, and second, that the computer is waiting for you to give it a command. If you were logged on drive A, you would see A>.

You will only see the drive specifier and prompt symbol when the computer is waiting for a command. After the computer has been given a command, the screen will usually change and new information may appear on it. After the computer is finished carrying out your instructions, the drive specifier will again appear, reminding you what your logged drive is, and the prompt will appear, notifying you that the computer is waiting for another command.

Files

So far we have discussed information contained on floppy or hard disks, or information manipulated in RAM. The problem arises of how to tell the computer which information you want to work with. The computer organizes information in files. You can think of computer files in the same way you think of information in a filing cabinet. A business letter may be stored in one manila folder, paid rent receipts in another, a contract in another, etc.

When using your computer, you will create files, and you will also use files you buy called *programs*. A program is a set of instructions for your computer, designed to accomplish a particular task. A computer program may consist of just one file, or a number of related files. Some examples of programs would be a word processing program to help manipulate words, a *paint program* used to manipulate visual images, a spreadsheet program which can be used for statistical analysis of business data, or a game program to help you wile away your free time.

Files need to be named so that you can tell your computer which file you want to use. Programs you buy come with a name. If the program is on a floppy disk, you would first insert the disk into the floppy disk drive, and then close the "door" to the drive. Let's say you placed a disk in drive A. If you happened to be logged onto drive C, you would type A: <cr>. (Computer books use <cr> to indicate when you should press the carriage return key on the computer's keyboard.) The computer will log onto drive A. You would then see A>. To *run* or start the program, you type the program's name into your computer, followed by a carriage return. The carriage return is necessary to let your computer know that you have finished typing in the name. The computer will search the logged disk drive for a file with that name. When it finds it, the computer will begin to follow the instructions contained in the program.

While using a program, you often will need to create a file. For example, when I start my word processing program to write a letter, the program asks me (by typing a question on my computer screen) what file I want to work with. Let's assume I type in DEARJOHN. My word processing program searches the disk drive I am currently logged onto for a file called DEARJOHN. If I have already created a file with that name, the program would *load* or display the file called DEARJOHN on my computer screen. If the file does not already exist, either because I have not previously created a file with that name, or because I had erased a preexisting DEARJOHN file, the word processor program will create a file named DEARJOHN, and will put into the file anything I type.

When I make a photographic image with my computer paint program, and decide to *save* it (take it from RAM and put it permanently onto a disk), I tell the paint program I want to keep the image by typing the appropriate instructions into my computer. The paint program asks me what name I want to give the image, that is, what name should be given to the file that will contain the picture. When I type in a descriptive name such as BEACH and push the carriage return key, the computer searches the current logged disk to see if that name already exists. If it doesn't, the computer creates a file with the name BEACH, *reads* the information from RAM, and *writes* it to the disk under the file name BEACH. (The information is now on a disk, but it is also in RAM. In effect, I have made a copy of the information that is in RAM, but I haven't destroyed it or altered it in any way.)

If the file name already exists on my logged disk, the program notifies me by typing a message on the computer screen, which may read something like, "This file already exists. Overwrite? Y/N." This message is informative and serves as a warning. The people who wrote the computer paint program are trying to protect me from myself. Each file must have a unique name. There would be little sense in allowing two different files to be called "BEACH," since the computer wouldn't be able to distinguish which file I was interested in. Which "beach"? So the program, having found that the file already exists, and knowing that I want to *write* a file with that name to disk, must destroy the previously existing file if I insist on using the same file name. If I type in y (short for "yes"), the computer replaces the existing BEACH file with the new one. If I type in n (short for "no"), the computer will ask for a new file name. I may decide to label the new photograph "BEACH2." If the computer does not find BEACH2 on the currently logged disk, it creates a file with that name, also on the currently logged disk.

Although you will be deciding what name to give your files, there are a number of rules that must be followed:

1. A file name cannot have more than eight digits and letters in it, although sometimes it can have a three-letter or -digit extension (more about extensions in a bit). Additionally, the file name may contain some special symbols which are located on the computer keyboard, but others are reserved for computer use, and hence are *illegal.* It's easiest to limit yourself to the 26 letters of the alphabet and digits. Although the file name cannot have more than eight characters/digits, it can have as few as one. Some examples of legal file names are: 1, MYLETTER, CHAPTER6, PHOTO127, p127, WORKING, NEWFILE, OLDFILE, and JILL. You will notice that I have typed the file names in UPPERCASE letters. When giving the name of a file, you can type it in uppercase or lowercase. To most computers, including IBMs and IBM-compatibles, uppercase and lowercase letters in a file name are the same. You will also notice that the names are not terribly descriptive, which is a product of the limitation of having only eight letters to work with.

2. The file extension which may be allowed is usually optional, and can be used to help make your file name more meaningful. An extension is a period (.) followed by a combination of up to three letters or digits. Some examples of legal file names with extensions are: Photo.1, AIRPLANE.502, and CAR.AD2. With some programs, user-defined extensions are not permitted, because the program needs to use certain file extensions to enable it to find information it needs to operate. Information such as this will be found in the manual that comes with the programs you buy.

3. When you tell your computer program that you wish it to open (use) a file, you usually must give the complete name. In my word processing program, if I had created a file called DEARJOHN.6, I would have to type the entire name, including the period and 6 into the computer to use that file. If I just typed in DEARJOHN, and the program could not find a file by that name, it would create a file called DEARJOHN. Computers are very literal. If you give one misinformation, it will either refuse to carry out the instruction, or do something you don't want it to do. Although file extensions may be helpful, they can just as easily be disregarded, and when first beginning computing, I recommend you limit yourself to the first eight (or fewer) letters/digits. If the computer program automatically appends a file extension (adds an extension to a file name you created), assume it is needed and meaningful for that program.

4. Finally, some file extensions are meaningful to the computer, and cannot be used. Never use the file extensions .COM or .EXE and avoid using the extension .BAT until you have gained some computer experience.

.COM and .EXE extension names are used by the computer to identify *programs.* Although programs are one or more files, they are special in that they give instructions directly to the computer. When you are working inside a program, you are not giving instructions directly to the computer, but using the program to translate your instructions into a form the computer can understand. The files you create when using a word processing program, or when using a paint program, are different from the files that comprise a computer program.

Let's say that you created a file in a paint program called PHOTO1, which has been saved on drive C. Sometime later you turn on your computer and type in the name of the file, in an attempt to put the image on the computer screen. At the C> you type: PHOTO1, followed by a carriage return. The computer will type an error message on the screen: BAD COMMAND. This happens because the file PHOTO1 is not in the proper form to be understood by the computer. You need first to go into the paint program you used to create the file PHOTO1, and tell that program to display the image on the screen. The paint program knows how to communicate with the computer, and your image will be displayed.

As described earlier, to get a paint program to *run,* all you have to do is type its name at the prompt. Let's assume that the paint program is called "PAINT.COM" and located on hard drive C. At the C> you type PAINT followed by a carriage return. The computer locates the file, determines that it is a program file, and starts to follow the instructions the program gives it. Notice that you do not have to type the extension .COM, although you can if you want. The same is true of files that end with the extension .EXE. The reason for this is because .COM and .EXE files are different from all other files, and the computer recognizes them as special, with special rules applying to them.

One very special type of program important to digital photography is called a *driver.* A driver is a program that acts like a translator between two other programs, or between a program and a piece of computer equipment. Usually a driver takes information from equipment and puts the information into a form the computer can understand, or takes information from the computer and puts it into a form the equipment can understand.

A Brief Review

By now you should be becoming familiar with the following concepts: Most of the work you will be doing with a computer will take place in RAM, the computer's electronic memory. Because RAM is not permanent, and information in it will disappear when the computer is shut off, information that you wish to keep is stored either on floppy disks, hard disks, or removable mass storage devices. Information that needs to be stored may be programs, which are special files that are capable of giving instructions directly to the computer, or files you create when using a program. Floppy disks have the advantage over fixed disks of being removable, so when one is filled up, a new one can be used. Because floppies are inexpensive, you can keep different types of information (files) separate from each other by grouping them on different disks. Business letters can be on one disk, photographs on another, etc. But floppy disks are limited in the number of bytes they can hold, and work more slowly than hard disks. Hard disks are fast and can hold large amounts of information, but once they are filled to capacity they either have to have files removed from them or a new hard disk has to be purchased. Removable mass storage devices have characteristics of both floppy disks and hard disks, in that they can hold relatively large quantities of information, they work relatively quickly, and they are removable so a new drive doesn't have to be purchased when the disk is full.

Add Ins

By providing space inside the computer cabinet, the capabilities of open architecture computers can be changed by the addition of new equipment. Such equipment may be made by the manufacturer, or by third party manufacturers. The *motherboard* (Figure 2.4) is found inside the cabinet of the computer. It is the main circuit board of the computer, and contains the *CPU* and other chips, as well as expansion slots which adaptor boards fit into. Figure 2.4 shows *expansion slots* on a motherboard inside an IBM compatible computer. These expansion slots are standardized, and act very much like an electric receptacle. In much the same way you plug a television into a wall outlet, you can plug an *adaptor board* (also in Figure 2.4) into an expansion slot. Adaptor boards are also called expansion boards.

To the untrained eye, most expansion boards look the same. They contain chips, specialized electronic parts, and usually some small switches which may

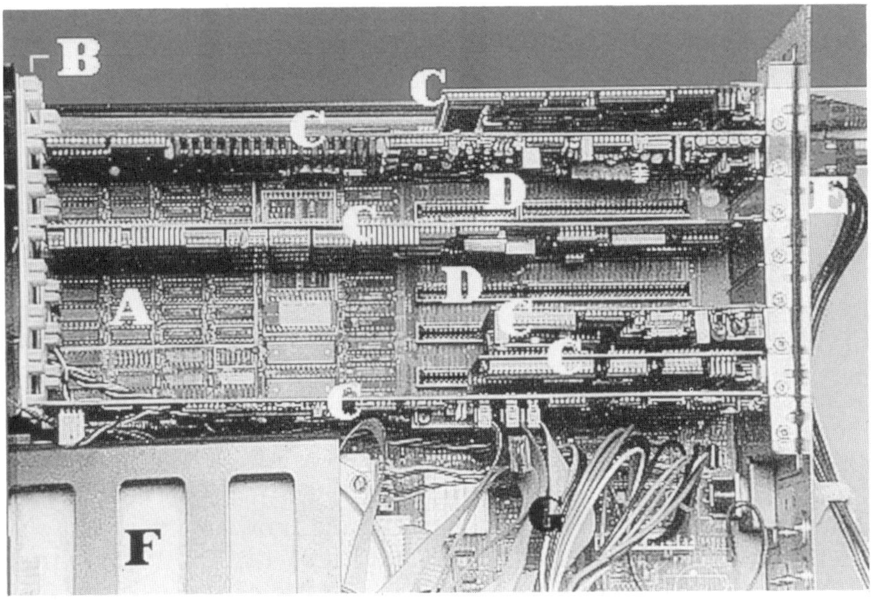

FIGURE 2.4 A motherboard and computer equipment: (A) computer chips; (B) guide slots used to help position and hold expansion boards properly; (C) expansion boards; (D) expansion slots that expansion boards plug into; (E) ports that cables plug into; (F) a hard disk; (G) miscellaneous wiring. Image made with TIPS software and utility programs.

have to be flipped (configured) on or off depending on what other boards may be inside the computer, in order to resolve compatibility problems. When a computer expansion board (often just called *board*) is purchased, it comes with *installation* instructions, and usually a telephone number to call if difficulties arise.

Expansion boards have been given that name because they expand the capabilities of open architecture computers. They are like attachments for a camera. You may purchase a bellows attachment for a 35mm camera to allow you to shoot close-ups, and you may purchase an expansion board to allow you to do sophisticated graphics (photography). We will discuss expansion boards in detail throughout this book.

Part of the motherboard is called the *bus.* The bus has sometimes been described as a highway, over which information from the expansion boards can travel to communicate with other boards. The bus is really a series of wires connecting all the expansion slots to each other, and to some chips in the computer. As we will see in Chapter 3, the speed at which the signals travel over the bus is important for digital photography purposes.

I/O

I/O (for *I*nput/*O*utput) refers to the transmission of information either to the computer or from the computer. This function is performed or controlled by the I/O expansion board. There are two different ways the computer can send or receive information. The first is called *parallel transmission*, and the second is called *serial transmission*. Parallel transmission sends one *byte* at a time from the computer to the peripheral, or from the peripheral to the computer. If you remember, a byte is made up of eight *bits*. To send eight bits at one time, each bit travels over a separate wire to or from the computer, simultaneously with the other seven bits. Therefore, a parallel cable is flat and wide, with wires running next (or parallel) to each other. Parallel transmission is the easiest for the computer to do, and often all the computer user must do to connect a parallel device is to connect one end of a cable to the peripheral device (e.g., something that is outside your computer) and the other to the end of the I/O board which sticks out of the back of the computer (Figure 2.4).

A serial device sends or receives information one bit at a time, similar to bullets being fired out of a machine gun. After eight bits have been received, one byte is assembled, and the next bits are sent and received. Because earlier computers used different systems for defining what a byte was, some computers used six bits to make a byte, others seven, eight, nine, or more. Today, a byte has been standardized at eight bits. But during the time when a byte could be anything from six to ten bits long, the computer needed to know how many bits made up a byte for the peripheral it was communicating with. Today, for purposes of compatibility, a personal computer still must be told how many bits make a byte when using serial transmission to communicate with peripheral devices. In addition, certain other information must be given to the computer to allow it to use serial transmission to communicate successfully with other devices.

Because parallel transmission is so much easier to set up than serial transmission, one may wonder why serial transmission would be used. Serial transmission does have at least one advantage over parallel. The cable from the computer to the peripheral can be longer, which means the device can be placed further away from the computer. This is convenient, and sometimes it may be necessary. The reason for this is that parallel transmission involves eight signals reaching their destination at exactly the same time. The longer the cable, the greater the chance that a glitch will occur in the transmission of one of the eight bits, and either create the wrong byte, or possibly, make the entire transmission erroneous.

Monitors

The computer screen I have referred to from time to time is more formally called a *monitor* (Figure 2.1), and can be thought of as a camera's viewfinder, since it allows you to see what is happening inside the computer, just as a viewfinder lets you see what is happening inside a camera.

Monitors may be monochrome, that is, they can display information in only one color plus black. Usually, monochrome monitors display black and amber, black and green, and less often, black and white. Monitors may also display information in color. Unlike color television sets, most color monitors are very limited in the number of colors they can display. Most can display either 4, 16, 64, or 256 different colors at one time. The limited color display results, in part, from the design of the monitor, but the most limiting factor is the graphics expansion board (usually just called *graphics board*) needed to run the monitor.

There are many different types of graphics boards, and we will discuss some of them in detail in Chapters 3 and 4, including some that display over 16,700,000 colors simultaneously. For now, it is sufficient to note that the amount of color displayed is dependent on both the design of the graphics board and the design of the monitor.

Besides the number of colors that can be displayed on a monitor at one time, the graphics board also governs the amount of pixels that can be displayed. The word *pixel* comes from the words *PIX* (short for "picture") and *EL*ement. A pixel is the smallest piece of information viewable on a computer monitor. If you look at a monitor (or TV) screen, you will notice that the image is made up of numerous dots. Each dot is a pixel. The number of pixels that can be displayed is called *screen resolution*. Some graphics boards are capable of displaying (sending to the monitor) a large number of pixels (high resolution), while others display relatively few pixels (low resolution). The greater the number of pixels, the more detailed the image, and the steeper the price.

One way to think of pixels is to imagine having a few thousand one-inch square colored tiles. Some are black, some white, some red, some blue, and some green. You can take the color tiles (pixels) and place them next to each other, and through the use of different colors, you can create a mosaic image. Since each color tile is the same size as every other tile, each tile contains as much information about the entire image as any other tile. This is what is meant by "a pixel is the smallest piece of information viewable on a computer screen." It should be obvious that the more pixels (color tiles) you have, the more detail can be included in the image. Also, the more colors available, the more realistic the resulting image.

The resolution of a computer's graphic system is not generally described by the number of pixels displayed on the monitor. If you look closely at a monitor screen, you will notice that the pixels are not randomly scattered about, but arranged in a rectangular grid. The lines of pixels that run from left to right are called *rows*, and the lines that run from top to bottom are called *columns*. Resolution is described by the number of columns displayed, by the number of rows displayed. For example, a low-resolution system may have a rectangular grid of 320 x 200 pixels, while a high resolution system may be 1024 x 768. When you see 1024 x 768, you can think, "There are 1024 pixels in each horizontal row and 768 pixels in each vertical column" or "There are 1024 columns by 768 rows." Since both are correct, this can get confusing for some (myself included). In this book, we will think of 1024 x 768 (or whatever numbers are given) as in the first description. Additionally, I will usually spell it out when discussing rows and columns. Figure 2.5 helps illustrate this point.

More About Memory

LIMS stands for *L*otus *I*ntel *M*icrosoft *S*tandard. As mentioned earlier, IBM and IBM-compatible computers were designed to be able to manipulate a maximum of 640KB of RAM, and I said that we would think of this RAM as *standard* RAM. Over the years the needs of computer users have changed. For example, digital photography was not practical in the early 1980s, because much needed

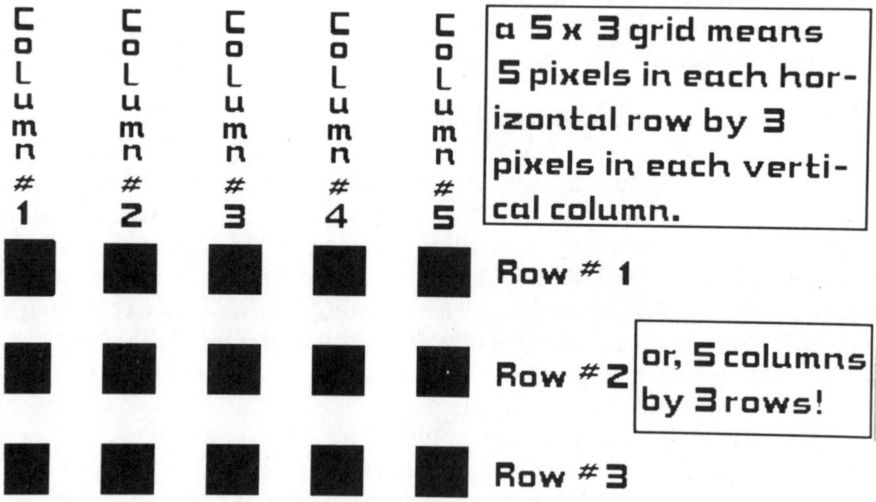

FIGURE 2.5

equipment did not exist. As this equipment was designed, it was found that the programs needed to manipulate digital photographs required more than 640KB of RAM to do a good job. Other programs, used for business and scientific research, also suffered from a lack of usable RAM to manipulate large amounts of information.

A number of the leading computer companies realized that some standard device should be developed to fool IBMs and IBM-compatibles into using more than the 640KB of RAM they were designed to use. If five different companies each developed their own method, there would be sales resistance to buying any of them. If you bought a device from one company, you would soon find that a program you wanted to use was made for use with a device made by a different company, using a different method to fool the computer. In effect, the "standard" IBM and IBM-compatible computers would start to become unstandardized. To stem numerous companies from devising their own methods, three giants in the PC computer world jointly agreed on a standard to be used, and made the specification available to any other company that wanted to manufacture the device. The three companies were Lotus, a manufacturer of computer software (another name for programs), Microsoft (another leading software and hardware manufacturer), and Intel, a leading manufacturer of computer chips and hardware (devices such as expansion boards). The device, which comes on an expansion board, was called LIMS.

LIMS are more often called EMS boards, and (usually) can be used in two different modes. The first *configuration* (method of *installing* hardware or software) is called EMS, which stands for *E*xtended *M*emory *S*tandard, and the second configuration mode is also called EMS, but this time for *E*xpanded *M*emory *S*tandard. Unfortunately, these two terms are difficult to distinguish from each other. *Expanded memory* is the method of fooling the computer into using more RAM to work with a program. On most computers we will be discussing for use in digital photography, you will be able to add anywhere from 512KB to 8MB of expanded memory.

Extended memory, on the other hand, is another type of disk. It is usually called a RAM disk, although the original name of *virtual disk* is still sometimes used. As the name RAM disk implies, it is an electronic disk, existing in RAM. It is super fast, because it works solely by manipulating positive and negative electric charges in RAM. Unfortunately, you usually have a choice of using LIMS/EMS *either* as expanded memory or extended memory, but not both at the same time. And of the two, expanded memory (fooling the computer into using more RAM to run a program) is more valuable in manipulating images than extended memory.

EMS expansion boards that adhere to the LIM standard can be fairly expensive. One Intel EMS board which comes with 512KB of RAM lists for over $500, and a fully populated board with the maximum amount of RAM permitted (usually 2MB per board) costs about $1,000. The LIM standard allows up to 8MB of expanded RAM to be used. Usually, one expanded memory board with 2MB of RAM on it will do for our digital photography purposes.

You should be aware that some alternative methods have been devised to use empty space on a hard disk as expanded memory. Most of these methods are implemented by specialized software (programs), which are very reasonably priced (usually around $100). However, these methods are not recommended, since they are usually *very slow* working and may make it difficult to actually get any work done, and some may cause the destruction of existing files (such as a photograph you have spent six hours working on) if you accidentally hit the wrong key on the computer keyboard.

Communications

Modems (from *Mo*dulate *Dem*odulate) are devices used by your computer to communicate over telephone lines (or via satellite with cellular phones) with other computers. Modems can either come as expansion boards, which fit inside your computer (internal modems), or as external modems, which reside outside the computer and attach to the computer by a cable.

Modems are necessary for telephone transmissions because the information from your computer must be changed into frequencies that can be sent over telephone lines. They also translate these frequencies back into a form the computer can understand. In the past, modems were used solely with computers, but they are now found inside every FAX (facsimile) machine.

How fast a modem can transmit or receive information is measured in a unit called *baud*. Most modems today run at either 1200 or 2400 baud, which is roughly equivalent to 150 or 300 characters per second, respectively. These speeds are much faster than the 300 baud speed of the first modems available for PCs in the early 1980s, and modems capable of 9600 baud, or 1200 characters per second, are now becoming common. The faster the modem is capable of communicating, the more expensive it is. There are 1200 baud modems available from about $70, and 2400 baud modems from about $120. Usually, 9600 baud modems start at about $600, and super-fast 19,000 baud modems at about $1,500. While modems are not necessary for digital photography, they can be helpful under some circumstances.

Software

We have discussed programs, which are specialized files that can communicate and give instructions directly to the computer. (As you probably remember, files you create when using a program cannot directly communicate with the computer, and need the program to act as interpreter.) Another name for programs is *software*, as distinguished from the physical components of a computer, expansion boards and peripherals, which are usually referred to as *hardware*.

Every computer needs specialized software specifically designed to make the computer operate. For IBMs this software is called DOS (for *D*isk *O*perating *S*ystem), and for IBM-compatibles, it is called MSDOS (for *M*icro*s*oft *D*isk *O*perating *S*ystem). Microsoft wrote the software that runs IBM's present line of personal computers, and sells the same software for use with IBM compatibles under the MSDOS name.

The name DOS implies that it has something to do with the disk drives, since DOS stands for disk operating system. Although DOS does contain programs that aid in disk operations, it also contains the information your computer needs to operate when it is first turned on, and contains other programs that you can use to tell your computer which I/O ports (input/output connections) should be used for different purposes, or to report to you how much standard memory is being used by a program.

PCDOS comes with IBM/PC/XT/AT computers, but MSDOS must be bought separately if you buy a clone. The series of programs come on a number of floppy disks, along with a manual that describes what each program does, and how to use it. There are numerous books sold that attempt to simplify and/or expand upon the original manual.

One problem that exists with DOS, and any other popular program, is that they get revised from time to time. While these revisions usually improve the program, they may entail some relearning. To distinguish one version of a program from another, a numbering system is used. The first version of a program may have the number 1.0 associated with it. For example, a hypothetical program designed to run a modem might be called "TALK, version 1.0" when it is first released (made available for sale). As minor improvements are made to the original "TALK" program, the version number is changed, usually in tenths. "TALK version 1.1" indicates that it contains some minor revisions from version 1.0. When a major revision in the program is made, usually to upgrade the program to take advantage of new technology or products on the market, or because some competing program has a lot of neat features TALK doesn't, the version number will be changed, but this time in whole numbers. A major change in "TALK, version 1.1" would make it "TALK, version 2.0."

When major revisions to a program are made, registered owners (e.g., those people who have returned the registration form included with the program) are usually notified and offered the new, upgraded version for a reduced fee. If the new version of "TALK, version 2.0" sells for $100, the registered owner may be offered it for $50. If you are the registered owner, and are happy with the version you have, there is no need to take advantage of this offer.

A program is copy protected if it cannot be copied. All manufacturers are concerned about *unauthorized* use of their programs. If you buy a program and make a copy of it and give it to a friend, you may be doing your friend a favor, but you are denying the manufacturer of the program a sale. To protect themselves from such unauthorized distribution of their copyrighted material, a few manufacturers employ sophisticated schemes to prevent you from making a copy of the program. This often means that you cannot put the program on your hard disk, or if the program will allow you to do so, you may have to keep the original disk the program came on in your floppy drive in order to be able to use the program from your hard disk. Sometimes these copy protected schemes are done through software manipulations, such as telling the program not to recognize the standard copy command if it is given, or through some physical means, such as physically damaging the disk with a laser beam at a critical point on the disk.

One major problem with copy protected software is that if the disk is damaged, the purchaser has to send the damaged disk, along with the original proof of sale and a nominal fee, to the manufacturer for a replacement. If the program is used for business purposes, the delay in getting a replacement disk could be very expensive, because of the *down time* involved.

Some companies make programs designed to *copy* copy protected programs. These programs usually work quite well, except with those programs that have been copy protected by physical damage to the disk.

When buying a program, make sure to ask whether it is copy protected or not. Many computer users refuse, on principle, to purchase copy protected programs, deciding to find a similar program, or go without the program, rather than "submit" to this process. Most companies do not copy protect their programs, believing that the inconvenience and bad will created will deter sales, and so resign themselves to the fact that a certain percentage of buyers of their programs will, in effect, "rip them off" by giving copies of the program to others.

If the program is *not* copy protected, it is not returnable. Programs come in a very well sealed outer envelope. Once the seal has been broken, the program is yours, whether you like the program or not. Human nature and ethics being what they are, there is nothing to prevent a person from buying a program, making a copy of it, and then returning the program, saying that he is dissatisfied with it, or doesn't like some of its features, etc. After all, one of the main functions of a computer is to make copies of files, and a program is just a special type of file. If this was permitted, most software companies would go out of business soon after going into business. You can return a *defective disk*, and the company will replace it. Otherwise, you're out of luck. In the computer software field, the saying, "Let the buyer beware," is literally true.

Keyboards

Up to this time I have mentioned keyboards, but have not described them. A computer keyboard differs from a typewriter keyboard in at least four ways.

First, the computer keyboard contains at least ten function keys, and sometimes more. Functions keys are labeled F1 to F10 (or higher), and are meant to be used with programs that have designated certain *functions* to these keys. For example, on one program, pressing F1 may bring a help menu onto the screen, prompting you what options are available (e.g., what to do next). On another program, pressing F1 may change the color of the letters on the screen, and on another, F1 may *exit* or leave the program. On still another program, the F1 key may do nothing. Information as to what the function keys do, or don't do, is included in the manual that comes with the program.

Second, the keyboard will (almost certainly) have a numeric keypad, similar to those found on calculators. This makes it easy to enter a lot of numbers, usually for business or statistical purposes. The numbers found on the top row of keys on a typewriter are also included on the computer keyboard. Most programs will accept *input* from both the numeric keypad or the more standard number keys, but some programs will only accept input from the standard number keys.

Third, the computer keyboard has *arrow* keys, used to move the cursor (a bar of light that shows where the next character will be printed on the screen, or where the next element in a drawing or photograph will be placed) rapidly from one place to another.

Finally, the computer keyboard has a number of other keys not found on a typewriter. For example, the *home* key will move the cursor to the top left corner of the monitor screen, and the *end* key will move the cursor to the bottom of the screen. Four other keys perform functions of special use in computer operations: The control key, usually abbreviated *Ctrl*; the *escape key*, usually abbreviated *Esc*; the *delete key*, usually abbreviated *Del*, and the *alternate key*, usually abbreviated *Alt*. As is the case with the *function keys*, these keys perform various functions, and their use is outlined in the computer's manual and in the manual that comes with a program.

SUMMARY

I have provided a great deal of information which may be a bit intimidating. The purpose of this chapter was not to instruct you in using a computer, but to give you a general idea of how computers work, and some of the things you will do with them. *Many* books have been written on basic computer use, and any of them will explain, in much greater detail, all of the concepts I have outlined in this chapter, as well as others I haven't. If you have never used a computer, it is doubtful whether you could now sit in front of one and begin to operate it. However, you should now be familiar with some of the basic concepts of computing.

For example, when you use a computer, you are actually manipulating positive and negative electric charges. Each positive or negative charge is called a bit, and a grouping of eight bits is called a byte. Each byte represents a character, digit, or punctuation mark. When you push the letter J on a computer keyboard, a program translates that computer key into a form the computer can understand. As you type information into the computer, it is put into RAM. It is in RAM that the information is manipulated. RAM is not permanent, but disappears when the electric current is turned off. To make information put into RAM permanent, it is stored on a physical disk. Some disks are removable, others are not. Some are capable of storing a lot of information, and others relatively little. Every personal computer contains at least one floppy disk drive, needed to put store-bought programs into RAM or onto a hard disk. Most computers also have at least one hard disk, because they can store a lot of information, and work faster than floppy disk drives.

Personal computers always associate themselves with a disk drive, and you can tell the computer which drive to associate itself with by typing the drive specifier associated with that disk drive on the computer keyboard, followed by a colon (:) and then a carriage return (<cr>). Besides a keyboard and disk drives, computers also contain a number of other devices. In a closed architecture computer, these devices are built in at the factory, while with an open architecture computer, they come in the form of expansion boards which you install yourself. These expansion boards are designed to be installed without specialized training or the use of specialized tools (a Phillips head screw driver may be necessary). Some of the more common expansion boards include one to control the computer monitor; one to hold extra RAM chips; one for I/O (input/output) needs, such as sending information to a printer; and a modem to send information through telephone lines.

When you use a program to write a letter or make a photograph, and want to save the letter or photograph on a disk, you create a file name consisting of one to eight letters and/or digits. No two file names can be the same on the same disk, because the computer could not distinguish between them if they did have the same name. With many programs you can also add an optional one- to three-letter/digit file extension name, which is preceded by a period, for a maximum of eleven letters and numbers. Some programs do not allow for optional file extensions, though, because the program must use specific file extensions to operate properly. Information such as this will be found in the user's manual that is included with the computer program.

If you understand the above summary, you are no longer computer illiterate. If someone should mention disk drive, expansion board, or pixel, you should not draw a blank. The chances are, though, that the meaning of some of these terms is still a bit fuzzy to you. If so, don't worry. Reread this chapter, and remember, you didn't pick up the jargon of photography, or whatever specialized area of knowledge you have, within the first few hours you were exposed to it. In the "Sources" section of this book are listed a number of good basic computer books that will help fill in the gaps, and explain in more detail the information I have sketched out. And remember, there are courses available to teach basic computer skills. To make digital photographs, all you will need are the basics. You won't have to become a computer whiz.

CHAPTER 3

BUYING A BASIC COMPUTER

In the previous chapter we learned some basic computer terms and concepts. In this chapter we will put some of that knowledge to work, to show how to construct a *base* computer. This base computer can be thought of as a camera body. In future chapters we will discuss adding the computer equivalent of film, lenses, and darkroom equipment, which you will need in order to use the computer as a camera.

The first part of this chapter is designed to help you differentiate between the IBM PC, XT, AT, and PS/2 line of computers. In the second part I will discuss computers that are compatible with these IBM systems. In the third part of this chapter the equipment a computer should have if it is to be used for digital photography will be discussed, and finally, I will suggest some base computer systems and methods for buying them. If you are familiar with this material, you may wish to skip to the end of the chapter where I will discuss specialized dealers of computer graphics and digital photography equipment.

This chapter will focus on IBM and IBM-compatible computers because they are open architecture computers, and thus there is an astonishingly large number of "attachments" available for them. This is not to say that other computers are not capable of being used for digital photography.

Recently, some manufacturers of digital photography equipment have begun designing equipment, or modifying equipment originally designed for IBM or IBM-compatibles, for use with the Macintosh II. While more equipment is available to run on IBM and IBM-compatible personal computers, the Macintosh II is catching up fast, and will soon offer as much in terms of power and flexibility as the IBMs and IBM-compatibles do. Because there are so many different types of IBMs and IBM-compatibles, we will concentrate on this family of computers. Since there are, or soon will be, equivalent hardware/software for the Macintosh II, most of the information in this book can also be applied to using the Macintosh II for digital photography.

Some computers, such as the Atari, Commodore, and others, are not adequate for the sophisticated types of digital photography this book presents. This is not to say that these computers cannot be used to make digital photographs. However, many third-party manufactures do not consider them powerful or flexible enough to warrant designing sophisticated hardware/software for them.

THE SAGA OF IBM

The IBM/PC (PC for *P*ersonal *C*omputer) was introduced in 1981, and over the next few years it went through a number of minor changes, although at the time these changes were viewed as major ones. The original model came with five expansion slots, no hard drive, and one 180KB single-sided 5.25 inch floppy disk drive—single-sided, because the floppy disk drive had one magnetic read/write head that looked for and put information on only one surface of the floppy disk. The original IBM/PC also came installed with 64KB of RAM, although more RAM was available as an optional upgrade.

Within a short time the original IBM/PC was upgraded with a 360KB double-sided 5.25 inch floppy disk drive. This drive had read/write heads positioned to read and write to both sides of the floppy disk. Both the 180KB and the 360KB drives were a standard size, now called *full height.* Third-party manufacturers soon started selling 360KB double-sided floppy drives, which were only half the physical height of the original IBM/PC drives, and the newer, smaller drives were dubbed *half-height* drives. Today, the standard size of a 5.25 inch floppy drive is the half-height. Because the original IBM/PC had enough room (called *bays*) for two full-height floppy drives, some computers can now hold four half-height floppy drives. Most newer computers have only three bays for floppies, because the size of today's cabinets (the outer box of the computer) is smaller than that of the original IBM PC's cabinet. This is because the "innards" of computers have been getting more compact as technology advances.

In 1983, IBM introduced the IBM/XT, an *upgraded* model of the original PC. The major innovation in the XT (which stands for e*X*tended *T*echnology) was an increased number of expansion slots from the PC's five to eight, and the inclusion of a 10MB hard disk as standard equipment. Because the PC was an open architecture computer, it was possible for a PC owner to install a hard disk drive in the PC, and so have the equivalent of an XT, but without the three additional expansion slots.

The increase in the number of expansion slots made the XT desirable, because the five slots of the PC were soon filled. One slot was needed for the floppy drive controller. Another slot was needed for a memory expansion board, since the *motherboard* of the PC did not provide receptacles for additional RAM. (The motherboard is the main computer board which contains both special chips to make the computer work, and expansion slots to accept expansion boards. Today, many IBM-compatible computers have receptacles on the motherboard for additional RAM chips to be plugged into, eliminating the need for a memory expansion board.) A third expansion slot was needed for the video controller board, needed to run the monitor. If the purpose of the computer was mainly for word processing, a monochrome monitor was usually used, because the quality of the first *color graphics controllers* (boards) was so poor that text (words) looked blurry on a color monitor. Many people had a need for, or just wanted, a color monitor for their computer, and so another slot was needed for a color graphics *adapter* (expansion board) as well as a monochrome adaptor. The fifth expansion slot was needed for an I/O (*I*nput/*O*utput) board, to allow the computer to send information to a printer, or receive information from other equipment. If a computer did use two different video boards (one for monochrome and one for color), there was no space for the hard disk controller needed to run a hard disk. The computer user had to decide whether to forgo a hard disk, or use only a monochrome or only a color monitor, because there were not enough expansion slots to get all the needed expansion boards into the computer.

Some third-party computer peripheral manufacturers introduced multifunction boards, which were two or more boards combined into one. For example, the I/O *ports* would be put on a monochrome expansion board or color graphics board, or an expansion board for RAM would also include a floppy drive controller, or a floppy drive controller and a hard drive controller were put on the same board. Buying a multifunction board freed one (or more) expansion slots, which could then be used for other needed equipment, and for new products that were being introduced into the computer market.

Even with the introduction of multifunction boards, IBM recognized that additional slots were necessary. To some extent, IBM realized that they were not in control over what their computers would be capable of doing. Although this bothered IBM, it was a major factor in making IBMs popular. Many small companies (some of which later became large companies), were identifying some specialized needs of computer users, and designing equipment to sell to these specialized markets. Providing more expansion slots became a marketing device for IBM. It became a way to get people who owned a PC, and who were frustrated because they needed additional slots to add needed features, to buy another IBM computer. Without the inclusion of extra expansion slots, the XT would just be a PC that came with a hard disk drive. With a number of companies making hard disk drives, there was no pressing need to buy one from IBM. Because of its open architecture, the cabinet containing the motherboard with its specialized, patented chips, was all that IBM had control over. What went into the expansion slots on the motherboard could be made and sold by anyone.

While third-party manufacturers were introducing new products that could be used in IBM/PC/XTs, computer enthusiasts were modifying the computers, often trying to "soup them up" to work faster. In the mid-1980s, articles started to appear in computer magazines describing methods to make IBM/PC/XTs run faster. The faster it could run, the faster it could accomplish a particular task.

Computers are designed to work at particular speeds. The speed at which the computer works is called its *clock speed*. The clock speed refers to the rate at which the computer seeks new instructions from the CPU (*C*entral *P*rocessing *U*nit). The IBM/PC/XT was designed to seek new instructions every 4.77 millionths of a second. Thus every second, the computer seeks instructions 4.77 million times. Because the speed of the computer can be thought of as how many times per second, or how frequently, it seeks instructions, clock speeds are denoted in terms of frequency. Thus the clock speed of the IBM/PC/XT was 4.77 megahertz (MHz).

The IBM/AT (for *A*dvanced *T*echnology) was introduced in 1984. Unlike the XT, the AT did represent a big leap in computer technology for the personal computer user. Like the XT, it too had eight expansion slots, but it came with a 20MB hard drive and a 1.2MB floppy drive instead of the 10MB hard drive and 360KB floppy drive of the XT. The change to a 1.2MB floppy drive caused some minor compatibility problems. The 1.2MB drive could read (get information from) a 360KB floppy disk, but could not *reliably* write (put information on) a 360KB floppy disk. The older PCs and XTs could not use information on floppy disks made by an AT. This incompatibility problem arose because the 360KB floppy disk has 40 *tracks* (circular magnetic rings containing information), while the 1.2MB floppy disk has 80 tracks. The solution to this incompatibility problem was to have both a 1.2MB and a 360KB drive on an AT computer, so information could be transferred to a 360KB disk and used on an older IBM/PC/XT. Adding a 360KB drive to an AT wasn't necessary unless the user had the need to use both types of computers, and hence was an option left to the buyer. Although the 1.2MB disk was a significant improvement in floppy storage capabilities, it was not available as an upgrade for the older IBM/PC/XT.

Many of the other improvements incorporated into the AT were of a technical nature, concerning how the computer manipulates information, and are beyond the scope of this discussion. However, the accumulated effect of these improvements was that the AT operated at a faster clock speed, increased from 4.77MHz to 6MHz. While arithmetically this appears to be a small increase, functionally it meant that the AT could perform tasks up to four times faster than a PC/XT. In 1986, IBM upgraded the AT to run at 8MHz, and included a 30 MB hard disk as standard equipment.

Because the AT was a completely redesigned computer, working on different computing principles than the PC/XT, it was heralded as potentially being able to do many things the older PC/XTs couldn't do. However, in an attempt to avoid alienating a few million IBM/PC/XT owners, the AT was made to be *downwardly compatible.* Software designed to operate on the PC/XT would run on the new AT, although much of the hardware (expansion boards) made for the PC/XT couldn't be used in the AT. Since many people had invested thousands of dollars in software, and many businesses had invested tens of thousands of dollars in software, this was a good marketing decision. Because the AT could use software designed for the PC and XT, potential sales resistance was overcome.

Computer software manufacturers had the option of creating programs that would take advantage of the increased technologic capabilities of the AT. To do this they either had to confine their programs to the AT, or they had to write two separate programs, one for the AT and the other for the PC/XT. To date, I know of few programs written specifically for the AT. Instead, software manufacturers have taken the easy route. Most still write programs as if for the PC/XT. Since these programs will also work on the AT, they eliminate marketing problems, such as having two different versions of the program, convincing software stores to buy and display different versions, having to write different manuals, etc. Although the AT has capabilities greater than the PC/XT, at this point the two classes of computers are very similar in terms of actual functioning, with the exception of the increased speed and increased floppy disk storage capacity of the AT.

Just as third-party hardware manufacturers had become important innovators in options available for the PC/XTs, they took a strong position in the AT market. Eventually it became apparent to IBM that they had completely lost control over the "options market." While a large and diverse add-in options market helped sales of the IBM cabinet, money for the options was going to third-party manufacturers, and not to IBM.

IBM introduced the PS/2 (for *P*ersonal *S*ystem) line of computers in 1987, and shortly afterward discontinued making the IBM/PC/XT/ATs. The IBM PS/2 models 25 and 30 were similar to the PC (but had only three expansion slots), the models 50 an 60 were similar to the AT, and the models 70 and 80 were similar to a more powerful class of computers just entering the market from a few other computer manufacturers. (The IBM/AT is also called a "286" computer, because it is designed around an Intel 80286 processing chip. The new class of computers are referred to as "386" computers because they are designed around the more powerful Intel 80386 processing chip. The IBM/PC/XTs were designed around the Intel 8088 processing chip, and are sometimes called "8088" computers.)

There are three main differences between the PS/2 computers and the ones they replaced. One difference is that the PS/2 includes, or in the case of the models 25 and 30 can be upgraded to, an improved high-resolution color graphics display (more pixels with more colors on the screen at one time). Within months of IBM's introduction of this new graphics system, third-party hardware manufacturers had introduced their own compatible versions to be used to upgrade IBM/PC/XT/ATs.

Also, the PS/2s have a new operating system, called OS/2 (for *O*perating *S*ystem), although DOS can also be used. Most PS/2 computer users have elected to stay with DOS, feeling that it is superior to OS/2.

The third difference lies in the *microchannel* expansion slots. With this innovation, IBM moved from an open architecture computer to a semi-open architecture computer. Unlike expansion slots on the PC/XT/AT, which can accept expansion boards from any manufacturer, the microchannel slots need a board with an electronic code embedded in it. A third-party manufacturer is still able to make a board for use in the new PS/2 computers, but they have to get a code number from IBM. This enables IBM to charge a royalty for each board a third-party manufacturer sells for use in a PS/2 computer. Thus, not only did IBM get some control over what goes into their computers, they also made money whenever an expansion board goes into their computer, even if another company designs, makes, and sells that board. For the computer user, the microchannel was said to have advantages over the standard expansion slots, because it was said to eliminate compatibility problems among different boards, and an attempt was made to make this into a selling point.

To help overcome some initial buying resistance from those who were not happy with the microchannel technology, IBM reintroduced the AT expansion system in a new model called the IBM PS/2 model 30/286, although this model only has three expansion slots.

IBM-COMPATIBLES (CLONES)

When the IBM/PC was first introduced, it went into competition with Apple and numerous small computer manufacturers marketing CP/M class computers. CP/M computers are now, for all practical purposes, extinct. They were nowhere near as powerful or flexible as the PC, and most were closed architecture machines. The rapid popularity of the PC created a growing market for PC clones. Almost all of these clones were not truly IBM-compatible, that is, not all programs designed to function on an IBM/PC would run on these computers. A more accurate term for these early computers would be *near-clones*.

The reason why these near-clones were not 100% IBM-compatible was that to be truly compatible, they would have had to violate a number of patents held by IBM, and in particular, one involving the BIOS, or *B*asic *I*nput *O*utput *S*ystem. The BIOS is that part of the computer that communicates with the keyboard, computer screen, printer, and other devices. Just how a computer goes about performing these tasks determines whether the computer is compatible with another computer. If one computer performs BIOS functions in one way, and another in a different way, then many computer programs written for one will not work with the other. Not many small computer companies wanted to risk a legal fight with the massive IBM. While it was possible for other companies to discover how the IBM BIOS operated, they couldn't legally duplicate it, so they designed their own BIOS. These very closely resembled IBM's BIOS, but close was not good enough.

For a few years IBM had virtually no competition (except for the Apple), since not enough near-clones were sold to make a significant dent in IBM's sales. If you wanted a true 100% IBM, you had to buy an IBM. The problem of making a non-patent-violating IBM-equivalent BIOS was overcome in the mid-1980s. While not all IBM-compatible computers on the market today are 100% IBM-compatible, almost all are. Two companies that make IBM-compatible BIOSs are Phoenix and Award. These companies sell their BIOS to third-party IBM clone manufacturers.

When IBM introduced the PS/2 line of computers with the microchannel expansion slots, ripples of fear spread over the computer clone and hardware manufacturing communities. They feared that the world would turn to the new microchannel system and they would be out of business, or at best have to pay royalties to IBM. But the world did not beat a path to the new PS/2s. Literally millions of IBM/PC/XT/ATs and compatibles were owned by this time, and users of these machines did not see why they should "scrap" their perfectly functional computers and buy new ones. One reason for this consumer loyalty was that many of the compatibles were better (an admittedly subjective opinion) and have more options (an objective fact) than the discontinued IBM models, and arguably are as good or better than the new line of PS/2s. Moreover, most compatibles cost one fifth to one third as much as the PS/2s.

The IBM PC, XT, and AT are no longer made. The clones of these computers have become so sophisticated that they have surpassed the computers they originally imitated. All PC class (8088) computers now come with eight expansion slots (at least I'm not aware of any that still only have five), and run at *both* the original 4.77MHz and a faster 10MHz clock speed. Although most software will run at 10MHz or faster, the original speed of 4.77MHz is retained as a safety feature in case a program is encountered that rebels at the faster speed. To switch from one speed to the other usually involves nothing more than pushing a button on the cabinet of the computer, although sometimes it is done by running a software command (e.g., typing one or two letters on the computer keyboard).

A similar situation exists with AT-compatibles. These computers have also become more than mere copies of the AT, most notably in the areas of clock speeds. Most are now available with either 10MHz, 12MHz, 16MHz, or higher clock speeds in addition to the lower 6MHz or 8MHz "safe" speed of the original IBM AT. As with the 8088s, these high and low speeds are usually hardware selectable, but sometimes they are software selected. A 10MHz AT-compatible monochrome system with a 20MB hard disk starts at about $1300, a 12MHz AT-compatible, similarly equipped, starts at about $1450, and a 16MHz AT-compatible begins at around $1800.

At this point we need to distinguish between the clock speed and the *bus* speed of the computer. The clock speed governs how fast the computer can look for and carry out instructions, and the bus speed governs how fast the expansion boards can communicate with each other. The bus speed on the later version of the IBM/AT was 8MHz, and because many expansion boards were designed to operate at that speed, changing the bus speed was not deemed a good idea. Today, most AT compatible computers maintain an 8MHz bus speed, although some have bus speeds of 10MHz or higher. As we will see in the next chapter, all of the different types of digital photography graphics boards you will have to choose from to manipulate your photographic images are designed to run at 8MHz. Any computer you get, regardless of its clock speed, should have a bus speed of 8MHz.

MINIMUM COMPUTER REQUIREMENTS

Because of the open architecture of the IBM/PC/XT/AT and now IBM-compatible computers, a great diversity of computer graphics equipment has been designed to run in these machines.

As we saw when discussing the saga of IBM, an open architecture computer really consists of nothing more than a box (cabinet) containing a motherboard. The motherboard has specialized chips such as the CPU (*C*entral *P*rocessing *U*nit), the expansion slots, a power supply which takes AC current and changes it to DC power, bays (room) for floppy and hard disks, and the all important BIOS (*B*asic *I*nput *O*utput *S*ystem), along with miscellaneous wiring. Most IBM-compatible computer manufacturers also include a host of other equipment, sometimes optional, and sometimes as part of a "package deal." Following is a checklist for what your computer system must have, along with some optional equipment you may wish to get. The list applies only to AT-compatible computers, since this is the type recommended by all manufacturers of digital photographic graphics boards. This is not to say that an XT cannot be used, only that it will put you at a disadvantage and so is not recommended. For one thing, the XT uses only 360K disks, and larger capacity (1.2MB) disks will be very helpful in making digital photographs. The difference in price between the standard XT-compatibles and the slowest speed AT-compatible is only a few hundred dollars, and this money will be money well spent.

Expansion Slots. Make sure the AT-compatible you select has eight expansion slots. Not only will this give you plenty of room for expanding your system, less than eight expansion slots may indicate a computer that is not as standardized as others that are available. There is no design reason why a motherboard should have less than eight expansion slots, unless (1) it's a truly off-brand computer, which may lead to problems in getting it repaired, or (2) it's a computer designed primarily for purposes not compatible with digital photography. For example, many *laptop* (portable) computers are designed for conducting business away from the office. These laptop computers are smaller than *desktop* computers and usually have only three expansion slots. While some laptop computers may be used for digital photography, a traditional desktop model will be more suitable.

Bus and Clock Speed. It is *extremely* important that the *bus* speed of the computer be 100% IBM-compatible. The bus speed must be 8MHz. The reason this is important is that many sophisticated computer graphics boards were designed for the IBM/AT with an 8MHz bus, and will only work at this speed. Although almost all AT-compatible computers have retained an 8MHz bus speed, some have not and have bus speeds of 10MHz or greater. Be sure to ask what the bus speed of the computer is, and if it is higher than 8MHz, don't buy it.

It is also important to make sure that the computer you choose has the proper *safe clock speed* setting of either 6MHz or 8MHz, in addition to any faster settings you may want, just to ensure that any programs you buy work properly. At this point you will have to make a decision: Should you buy a very fast computer for digital photography purposes, or should you buy the slowest one on the market, or one of moderate speed? I recommend you get the slowest computer you can find, which will probably be 12MHz (clock speed). An increase in clock speed will not mean much in digital photography, although there are probably instances where it will save you some time. For the extra money, though, it really isn't worth it. I use an older computer with a top clock speed of only 8MHz, and I spend most of my digital photography time looking at the image on the screen, deciding what to do next. This type of computer use does not require a speedy computer. Moreover, when those few functions that do take 45 seconds are encountered, you can stand up and stretch.

Buying a faster (clock speed) computer will not cause problems when used for digital photography. The only point is money. A computer with a 16MHz clock speed may cost 33% more than a 12MHz computer, and a 20MHz computer may cost 80% more than one running at 12MHz.

The computer industry doesn't think as I do, and speed is of paramount importance to them. So much so that 8MHz AT-compatibles are no longer made, being deemed too slow. If you intend to use a computer to do a lot of business-related or statistical computing, involving manipulating a lot of numbers, or if a number of people are going to use the computer, then consider a faster clock speed computer, since the time it saves can make a difference in these cases. Otherwise, save your money.

RAM. The amount of RAM that is included in your system is another factor. Most AT-compatible computers come with 512KB of standard RAM, but you should buy an additional 512KB and have it put on the motherboard when you buy the computer. One reason for this is that many programs require 640KB of standard RAM. It is possible to buy an additional 128KB of standard RAM and add it to the 512KB most AT-compatibles are sold with, for a total of 640KB. This is a bit less expensive in the short run, but you will eventually need 1MB of RAM. This will be discussed again later in this chapter.

Bays. The amount of bays, or receptacles for floppy drives, may be a factor. Since most drives are *half-height*, and since the original IBM/ATs were designed to hold two *full-height* drives, some computer cabinets are equipped with bays capable of holding four floppy disk drives. But as mentioned earlier, most now hold three half heights. A computer cabinet that is capable of holding three floppy drives should be sufficient, since many users will start (and possibly stay) with only one floppy.

Floppy disk drives. For AT-compatible computers, there is usually a choice of getting either a 5.25 inch 1.2MB disk drive or a 3.5 inch 1.44MB disk drive. Although the 3.5 inch 1.44MB disk can store a little more information, the first ATs came with the 1.2MB disk drive, and the industry standardized with disks of this capacity. The 3.5 inch disk was introduced with the PS/2 lines of computers, and soon hardware manufacturers adapted it for use in AT-compatible computers. Although the 3.5 inch disk is becoming more popular, most AT-compatible computers have a 1.2MB floppy disk drive. Moreover, most programs come on a 5.25 inch 360KB disk, which can be read by a 5.25 inch 1.2MB disk drive, but not by a 3.5 inch 1.44MB disk drive. For these reasons, I suggest that when you buy the computer, you request the 1.2MB floppy disk drive, instead of the 1.44MB. The cost of the computer usually includes one floppy drive, and the price is usually the same regardless of the one you choose. If you determine at a later date that you need a 1.44MB drive, you can buy one and add it to your system. An additional drive of either type usually runs between $80 and $120.

Hard disk drives. While the floppy disk drives have slots or doors that open in the front of the computer cabinet to allow the user to insert and remove the floppy disks, this is not the case for hard disk drives. There is no reason for these drives to be easily accessible because the disks are fixed into position and are not removable. Inside the computer cabinet there is usually space (bays) for two half-height (or one full-height) hard disk drives. Hard drives with capacities of 65MB or less are usually half height, but not always, and larger capacity drives are almost always full height.

Although the inclusion of a hard disk drive is usually offered as an option, it's really a necessity. Manufacturers often advertise the price of their computers without the inclusion of the hard disk to keep the initial cost down. I can think of no computer applications, with the possible exception of playing some games, that do not benefit significantly from the use of a hard disk. Some computer software only works with a hard disk, and "hard disk needed" will be printed somewhere on their packaging or in their advertising. For a program that will work on both floppy and hard disks, the manual accompanying the software almost always includes the notice that the program will work better (e.g., more quickly and easily) on a hard disk.

Because digital photography involves making image files that can easily get larger than the storage capacity of a 1.2MB floppy disk, and because many computer software programs needed to make digital photographs run 500KB to 1MB in size, a hard disk is necessary to store and manage these files and programs.

So the question becomes not whether you should have a hard disk or not, but what storage capacity you should buy. The minimum-capacity hard disk for AT-compatibles is usually 20MB. A hard disk of this size is sufficient to begin with, although you may find after a while that a larger hard disk will be necessary. If you plan to use your computer for other functions in addition to digital photography, a 40MB hard disk is advisable. Other programs, such as word processors, may use as little as 256KB, or as much as 1MB of disk space. Most utility programs take up 32KB to 500KB of space. You may find that as you become aware of the diversity of programs available, you will begin to fill up your hard disk with numerous word processing programs, bookkeeping programs, and utility programs, not to mention programs to help make digital photographs. For this reason, unless cost is a major factor, a 40MB disk will be your best buy.

There are minor price differences among drives of the same storage capacity, mainly due to the different marketing strategies of the competing manufacturers. The careful buyer will also notice that two hard drives with the same storage capacity made by the same manufacturer may have different prices. This is because the *access times* the drives work at are different. Access time refers to the average time it takes the read/write head to move from track to track when seeking information. The differences in access times are usually in the order of a few milliseconds (thousandths of a second) and a slower drive (longer access time) will not put you at a functional disadvantage, especially if you are using a computer with a (relatively) slow clock speed. Even a "slow" hard disk is lightening fast when compared to a floppy drive. One thing most computer "speed freaks" don't take into consideration is the human element. If a faster computer finishes a task a few seconds before a slower one, the time saved will be lost if the telephone rings and you answer it, or if you sneeze and look for a tissue. Fast computers can save *real* time when doing such tasks as statistical analyses with very large data sets, or when doing many computations with large *spread sheets* for business purposes, (e.g., for accounting or forecasting next year's sales revenue). For most computer applications, such as word processing and almost all digital photography, great speed just isn't needed. So if cost is a factor, buying a hard disk with a slower access time will not put you at a disadvantage compared to a hard disk with a faster access time. A 20MB hard disk can cost as little as $250, 40MB hard disks are priced as low as $400, and an 80MB hard disk starts at about $650. Probably the most widely used hard disk drives are made by Seagate, but drives by MiniScribe, Maxtor, Mountain, and CDC (among others) are also popular.

Earlier I mentioned that most computers today have room for only two half-height hard disks. Large-capacity disk drives, those over 65MBs, often come only in full-height sizes. If you have a half-height 40MB drive, and at some future time want to add a full-height 80MB drive, you will probably find that there is no room inside the computer for it. The solution to this problem is to buy an *external cabinet* for about $125 to $150 for the additional hard disk drive. As the name implies, this cabinet or box houses the drive and sits outside the computer. It is connected to one of the computer's I/O ports by a cable.

Ports. Input/Output ports allow your computer to work with peripheral devices, such as a computer printer, some types of modems, a hard disk external cabinet, and equipment you may need for digital photography. You can get I/O ports in the form of an expansion board, which usually costs between $50 and $75. A typical board will include one serial port and one parallel port, which protrude from the back of the computer. To connect the peripheral to the computer, a cable is used. One end of the cable plugs into a port (receptacle) of the I/O board and the other plugs into a port built into the peripheral.

Serial ports come in two main varieties, called RS232 and RS232C. The RS232 port has 25 small female receptacles, while the RS232C has only 9. Most of the 25 receptacles of the RS232 usually aren't used, and so the 9-receptacle version was introduced to save manufacturing time and materials. It may turn out that a serial peripheral device has one type of port, and your computer has the other. This causes an incompatibility problem which can be solved by buying a converter to change one of the ports to the other type. These converters are similar to those you might use to convert a three-prong plug on a kitchen appliance to fit a two-prong wall outlet. These converters cost less than $10.

Although you need at least one serial and one parallel port in your computer, and although I recommend buying the ports on an I/O expansion board, there are some other options available to you. Many other boards, most notably standard graphics boards, often come with a serial and parallel port either as standard equipment or optional equipment for an additional fee of about $35. This *bundling* of I/O ports with graphics boards is a holdover from the earlier days when there were only five expansion slots in computer cabinets. You can elect to pass on buying the I/O expansion board if your standard graphics board comes with ports, although you will find that as you start buying more peripheral devices, extra ports are necessary. Moreover, on some AT clones both a serial and a parallel I/O port are built into the motherboard. If this is so in the computer you purchase, you still may want to get additional ports either on a standard graphics board or on an I/O board.

Standard graphics expansion boards. The image you see on the computer monitor is generated by the graphics expansion board. To a large extent, standard (non-digital photography) monitors and standard (non-digital photography) graphics boards are a matched set. In this section we will discuss standard graphics boards, and in the next section we will discuss standard monitors.

When the IBM/PC was first introduced, the buyer was given the choice of purchasing a *monochrome board* which only displayed text (words) but not graphics, or a *color board* which was capable of displaying both text and graphics, such as charts and games, in color. The problem, as mentioned earlier, was that the quality of the color graphics display was so poor that it was difficult to read text on the monitor, so buying just the color graphics board to do both word processing and graphics was not a good solution for many users. People who needed both graphics and text often had to buy both a monochrome board and a color board (they also had to purchase both a monochrome monitor and a color monitor). A third-party hardware manufacturer called *Hercules* introduced a system that allowed both text and color graphics to be displayed on a monochrome monitor. However, the graphics were displayed in a limited number of monochrome shades (e.g., shades of black and amber, black and green, or black and white, depending on the type of monochrome monitor being used). Today, almost all monochrome boards are "Hercules-compatible," which means that a monochrome *adaptor* (board) will allow you to run many, but not all, color programs with a monochrome monitor. For the most part, a computer with a standard monochrome board will only be limited as far as the number of games that can be played on it.

In time, IBM introduced a number of improved color graphics *standards*, each able to display more pixels and more colors on the color monitor. Within a few months of the introduction of each new "state of the art" standard, third-party manufacturers introduced their own versions. To upgrade to the new standard, the consumer had to purchase a new graphics board. These boards were always downwardly compatible, that is, they could run software programs designed for the previous standards as well as the new standard. Not only was the hardware designed for maximum compatibility, but software manufacturers were often able to design their new programs so that they would also work with the new standard as well as with the older standards. In this way, regardless of whether the consumer decided to upgrade or not, she could still buy many new or improved products designed for the new standard.

The earliest color standard, CGA (for *C*olor *G*raphics *A*daptor), is still available today, although it is close to becoming extinct. The second standard, EGA (for *E*nhanced *G*raphics *A*daptor), is still alive and doing well. The next graphics upgrade was VGA (for *V*ideo *G*raphics *A*rray), and is still the standard, although a newer version, Super VGA (abbreviated *SVGA* and sometimes called *enhanced VGA*) is available. Super VGA has not become the standard yet, because its improvement over VGA in terms of pixels and colors is not considered tremendous. Many software manufacturers do not design programs to run with Super VGA, although Super VGA graphics boards will run both Super VGA, VGA, EGA, and often CGA graphics. As more people upgrade to Super VGA, more software manufacturers will design programs to take advantage of its increased abilities. Apparently, many computer users are reluctant to spend the money to upgrade to Super VGA and are waiting to see what new *graphics standard* will materialize. (New graphics standards seem to be introduced to the market every two years or so.)

With each succeeding color graphics advance, the clarity of text displayed on the computer monitor increased. Because of this, many computers today use a color (EGA/VGA) graphics board. This brings us to another decision you will have to make: whether to have a monochrome or color standard computer board and monitor. Since this is a book about digital photography, you may think that this is an easy decision to make; if you want to make color pictures, you may think you need a color EGA, VGA, or Super VGA adaptor. But almost all of the color graphics expansion boards we will be discussing for use in digital photography are highly specialized, and go well beyond the capabilities of even Super VGA. Unfortunately, with a few notable exceptions, these specialized boards will not run programs designed for use with EGA, VGA, or Super VGA expansion boards. This means that the specialized digital photography boards are only useful for making digital photographs (as well as animation, sophisticated drawing and painting, and to some extent, for computer-assisted design and drafting, called CADD). They cannot be used for other computer needs. Even if you want to limit your computing to digital photography, you will need a *standard* graphics adaptor for such mundane but necessary functions as reading messages from DOS, for file management, making copies of files, changing disk drives, etc.

Therefore, you will have to make the following decision: In addition to the highly specialized graphics board you will probably be using to make digital photographs, should you also have a standard color graphics board in your computer or a standard monochrome board? If you choose the monochrome board, you will save *a lot* of money, which may be better spent on digital photography equipment, and you will be able to run many programs designed to run on a color board, but not all. If you choose a color graphics board, it will cost more money, but you will be able to run all programs designed for use with PC/XT and AT-compatible computers.

There is no easy solution to this problem. I have decided to use a monochrome board, because even the best (Super VGA) graphics expansion boards produce images that I find esthetically unacceptable. I'd rather look at type produced by a monochrome board, and look at high-quality color graphics produced by my digital photography graphics board, than look at the rather "cartooney" images produced by the current standard color graphics boards presently available. However, I am in a minority, since most users of sophisticated color graphics equipment also use EGA or VGA color for their non-digital photography needs.

Monitors. A standard color graphics system is more expensive than a standard monochrome system, because color requires a color monitor, which is usually more expensive than color televisions of the same screen size. Additionally, not just any color monitor will do; you need one that is matched to the *frequency* of the graphics board you choose. By frequency, I mean the electronic signal that the board uses to send information to the monitor. As mentioned earlier, a graphics board and graphics monitor are a matched set.

Standard monochrome boards, also called *monochrome display adaptors* (MDA) usually start at about $60, and monochrome monitors usually cost between $90 and $120. A monochrome display board and monitor can usually be purchased together for as little as $130. Almost all monochrome display boards will work with any monochrome monitor, since the frequencies of each have been standardized.

Unfortunately, color graphics expansion boards have not been standardized to emit signals at the same frequency. A general rule is that the higher the quality of the board (the more pixels and colors the board is capable of displaying), the higher the frequency of the video signals sent from the board to the monitor. The monitor must be able to receive the same frequency that the board emits. If there is a mismatch, the color graphics system will not work.

The lack of standardization in color boards and monitors became troublesome to many computer users. With the introduction of each new and improved graphics standard, a new monitor was required. An upgrade to VGA from EGA would entail not only buying a new graphics board, but also a new graphics monitor. VGA boards put out a higher frequency signal than did the EGA boards and EGA monitors cannot receive the higher VGA signal.

Some monitor manufacturers solved the incompatibility problem, at least partially, by introducing multisync monitors. These monitors are capable of receiving a wide range of frequencies, instead of being limited to one specific frequency, as was the case before the introduction of the multisync monitors. One monitor could be run with MDA, EGA, VGA, Super VGA, and, theoretically, with new "standards" not yet devised. The multisync color monitor is (usually) capable of automatically determining the frequency of the signal emitted from the color graphics board and adjusting itself to that signal. This is handy if you have elected to have both a monochrome display adaptor for text processing, and a color graphics adaptor for EGA or VGA graphics. In such a case, you can tell the computer which graphics board to use, the computer switches to that board, and the multisync monitor automatically adjusts itself to the frequency being sent by that board. Most multisync monitors, and other "high end" monitors, have a *text switch* or button on their outer cabinet that enables viewing text in either black and green, black and amber, or black and white, which supposedly makes the text look sharper and clearer then when viewed in the full color mode.

The cost of standard color graphics systems (a color board and color monitor) varies greatly. A low quality CGA graphics board and monitor costs about $275, although it is becoming more and more difficult to find CGA systems. A higher quality EGA system with a 13-inch average quality monitor can cost as little as

$475 , but a better quality 13-inch monitor can increase the cost to $750. A still higher quality VGA system with a 14-inch average quality monitor usually starts at about $700, and selecting an enhanced (higher quality) monitor can bring the price up to $1000. Multisync monitors, without a graphics board, usually can be found for $700 to $1200, and VGA boards, without a monitor, can be found starting at about $250. A "state of the art" color graphics system using a multisync monitor and quality VGA board will cost at least $1000, and probably more.

There are a few vocabulary terms specific to color monitors which you should be aware of. First, the word *monitor* itself has a meaning. A monitor consists of a cathode ray tube, similar to the one in a television set. The main difference between a monitor and a television is that the TV has a *tuner* which allows different channels to appear on the screen. Monitors do not have a tuner. Because a TV set and a monitor are so similar, it is possible to use a TV as a monitor. This usually necessitates using an *rf modulator*, which functionally turns off the TV's turner circuitry. Using a TV as a monitor is not recommended. The quality of the image is poor, and it may be difficult to get the computer display centered on the screen, or to keep it from rolling.

Monitors used for standard computer applications (that is, not for digital photography), use short persistence phosphors. The image from a monitor (or TV set) is made up of electrons being sent (fired) out of a "gun" and hitting the back side of the monitor screen. These electrons excite portions of a phosphorescent coating on the back of the screen. The phosphorescent coating glows when hit by the electrons. Short persistence phosphors glow for a brief period of time. Long persistence phosphors glow for a longer period of time. Short persistence phosphors are used for most conventional computer applications because long persistence phosphors tend to cause a "ghosting" image. As the image on the monitor changes, a slight afterglow remains of the previous image for about a second. Many people find this annoying.

The earliest color graphics adaptors had just the opposite problem. The image *flickered* on the screen. Advances in computer graphics and the design of monitors have effectively eliminated the problem of flicker (unless you are using a CGA system). Unfortunately, *short persistence phosphors* are not adequate for use in many digital photography applications. The *flicker* remains in these systems, and can make using monitors using short persistence phosphors for digital photographic purposes self-defeating, because the image information displayed on the monitor from the program is difficult to view. For most digital photography purposes, monitors with long persistence phosphors are needed. These monitors are not easy to find; they are usually sold either directly by the manufacturer or through a select group of authorized dealers.

The short persistence monitor you choose for the standard computer applications will not serve you well when used for digital photography, so you will in all likelihood need to get a second (long persistence) monitor specifically designed to work with your digital photography graphics board. The investment of $1,000 or more for a quality multisync monitor and quality graphics board will almost equal the present (low) cost of the long persistence monitor needed for digital photography. These monitors usually sell for $1,200 to $2,500. One consolation is that most of these long persistence monitors can be made to work with a variety of VGA/EGA boards, usually requiring nothing more than the addition of a specialized cable. One very safe solution is to start off with a monochrome system for your regular computing needs. After using it for a while, decide whether you need a color display for your non-digital photography needs. If so, an appropriate EGA/VGA board can be purchased and used with your long persistence digital photography monitor, or the investment can be made for a standard short persistence monitor.

Expanded memory. For many digital photography manipulations, *expanded memory* will be needed. Expanded memory fools the computer into using more than the 640KB of RAM it was designed to use to manipulate information. This is why earlier in this chapter I recommended that you have 1MB of RAM installed on the motherboard of your computer. In order for your computer to use expanded RAM, it must have 1MB of RAM already in it. Think of it this way: The first 640KB of RAM is used by your computer to manipulate information. The additional 360KB of RAM prepares the computer to use expanded memory. 640KB + 360KB = 1MB. RAM above 1MB can be used for expanded memory.

It is possible to get involved in digital photography, and even do rather sophisticated work, without the expanded memory, but eventually you will become frustrated that there are some things you will not be able to do and some programs you will not be able to use. A number of expanded memory boards are available, although there is some evidence that few are actually 100% compatible with the LIM standards set by Lotus, Intel, and Microsoft. Since programs are written to work with LIMS specifications, the lower cost of many of these boards may not be a true bargain. The Intel *AboveBoard*, which does meet the LIM standard, is expensive. The board itself, which comes with 512KB of RAM installed (already on it), lists for about $500. Most programs that require expanded memory recommend a minimum of 1.5MB or 2MB, which may raise the price to about $750. (You should also be aware that the *AboveBoard PS* includes one parallel and one serial port for about $50 more than the *AboveBoard*, so this board is a good way to add extra ports to your computer without having to use extra expansion slots.)

Modems. While a modem will allow you to use your computer to communicate with other computers, modems have only limited usefulness in digital photography. In later chapters we will discuss service bureaus, which are the digital photographer's equivalent of a photo lab. A modem would allow you to send a file containing a digital photograph from your computer through telephone lines to the service bureau, but the cost of transmission can get expensive. For example, it could take over an hour to send a 500KB file (one that represents a not particularly complex digital photograph) with a 1200 *baud* modem. Assuming that the transmission is flawless, this may represent an acceptable price to pay (one hour of probably long-distance telephone time), but if there is electronic noise or static on the telephone line, or if a millisecond glitch in your modem or the receiving modem occurs, the resulting photographic image may be ruined.

One company, Telebit, sells a modem that has an extremely fast 19,000 baud transmission speed and is said to be 100% accurate in transmission quality. But instead of the usual $50 to $300 cost of most modems, this product costs over $1,500. Furthermore, for this product to be useful the service bureau must also have a modem capable of receiving such a rapid transmission. You can't send fast and have them receive slow. This option is mentioned in the event that you find that you really need to send files to a service bureau by modem; however, for most people overnight mail or Federal Express will be more cost effective and functionally as fast, as we will see when discussing service bureaus.

Keyboards. There are a number of different types of computer keyboards, and although manufacturers will disagree, there really aren't great differences among them. All computers come with a keyboard, and sometimes you may be given a choice among a number of different ones. If you can't decide, just ask for the most popular model.

BUYING YOUR COMPUTER SYSTEM

All of the computer equipment described in this book can be put into one of two categories: that available at local computer stores and by mail order, and equipment that is sold through a limited number of specialized or franchised dealers or directly from the manufacturer. The following section describes three methods you can use to put together your basic computer system for digital photography. The assumption is that you are not familiar with purchasing computers, installing expansion boards, or getting programs onto the hard disk of the compu-ter. Although the first two methods described below may not be of vital interest to those readers who have such experience, the third section will be, so don't skip it.

Computer Stores

A digital photography system requires a basic computer system including the equipment described in this chapter. In addition to this *base system*, you will add specialized equipment. Standard computer salespeople know nothing about the specialized equipment you will need, and will probably try to talk you into buying their "desktop publishing system" which is "on sale for only one more day" and "will do everything any other system will do, but better." I realize that generalizing about all computer salespersons is dangerous, since *some* must be knowledgeable. It seems to me that today, most clerks in computer stores have a very limited knowledge of computers, and no knowledge of digital photography. Of the three methods for buying a computer described in this chapter, buying one from a computer store is the one to try to avoid at all costs. Unfortunately, it will appear to many as the easiest way to go. It isn't.

All computer stores will assure you that they will be available to offer assistance when problems arise. Most computer stores will answer a few questions for a few days after they sell the equipment, but then they usually become less and less available. There is no profit in holding your hand. As for repeat business, the turnover at computer stores is so great that many salespeople realize that they possibly won't be working there when, and if, you come back for another product.

The cost of equipment at a computer store is usually at or very close to list price, although some discount stores do exist that give legitimate discounts. Unfortunately, almost all stores claim to be "discount computer stores." But even discounted prices are higher than mail order. Stores try to sell only what they have in stock, or what they can easily order for you. To get you to buy these items, they often use interesting tactics. "If you don't buy THIS computer you're wasting your money. What computer were you interested in? THAT? Why, THEY have close to a 100% failure rate right out of the factory. (I've actually been told this.) THIS is the only trouble-free computer made. THAT'S why WE sell it. WE don't sell junk."

The only things you should buy at a computer store are those that cost $20 or less: a box of floppy disks or a cable. (If you buy a cable at a computer store, make sure you can return it if it is the wrong type. Don't go by the assurance of the salesperson that it is the type you need.) In the previous chapter I did mention that some computer stores give basic computer instruction. This could be another valid reason for venturing into a computer store, but if you do, be on guard for subtle, or not so subtle, sales pitches.

Mail Order

Mail order is reported to account for over 70% of all computer, hardware, and software sales. Mail order prices are significantly lower than prices at computer stores, and as there is strong competition among the various mail order companies, prices tend to stay down. As far as *generic* computers go, that is, computers bought in parts by the mail order houses and then assembled by them, the final products look similar if not identical to each other in ads, and their specifications are also generally similar. Buying from one major mail order house is much the same as buying from another. Generic computers *tend* to be extremely IBM-compatible, and compatible with each other. Standard parts are bought from manufacturers who design their equipment to easily work with other computer equipment.

Once you have gained some experience using computers and installing hardware, you may turn to mail order as your usual means of buying standard equipment. Buying your first computer by mail order can, for most people, be somewhat intimidating. What should I buy? How do I know they are honest? What if it doesn't work? About once a year major computer magazines, such as *PC*, *Personal Computing*, *Byte*, and some others evaluate computers, some from "name" manufacturers and some from mail order houses. The companies listed in the "Sources" section at the back of this book under Mail Order Houses are ones whose products have received good reviews in recent years. While this will not guarantee that these companies are still producing outstanding computers, it does indicate that at least they are capable of doing so. Some of these companies only sell their own computers by mail, and others also sell a variety of products from other manufacturers.

If you buy from a mail order house, this is what you will need.

Base system 1. This is a moderate-cost system. You need:

A. A 100% IBM/AT-compatible ("286") computer: one 5.25 inch 1.2MB floppy drive, enough bays to hold three half-height floppy disk drives, a combination hard drive and floppy drive controller, 512KB of RAM, and eight expansion slots, with a Phoenix or Award BIOS. I recommend a maximum clock speed of 12MHz or 16MHz, with a safe clock speed of either 6MHz or 8MHz. IT MUST HAVE A BUS SPEED OF 8MHz. Current cost about $950.

B. An additional 512KB of RAM placed on the motherboard. Price about $250.

C. A Hercules-compatible monographics board with a monochrome monitor. Cost about $130.

D. An I/O board with at least one parallel and one serial port. Cost about $60.

E. A 20MB Seagate or equivalent hard disk, with controller if the base system does not come with a combination floppy and hard disk controller. Cost about $225.

F. MSDOS, version 3.2 or 3.3. Current price about $80.

G. A keyboard will be included in the price of the computer. If given a choice, ask for their most popular keyboard.

H. All appropriate cables, with all equipment installed and the system completely set up and ready to run. This should be included in the overall price.

Total price of Base System 1: about $1700 plus shipping.

Base system 2. This system is the same as Base System 1, but with the following enhancements:

A. For a 40MB Seagate or equivalent hard disk instead of the 20MB hard disk, add about $100.

B. For an 80MB Seagate or equivalent hard disk instead of the 20MB hard disk, add about $475 to the base system.

C. For a 3.5 inch 1.44MB floppy drive *or* a 5.25 inch 360KB floppy drive in addition to the 5.25 inch 1.2MB floppy drive that comes with the computer, add about $100.

D. An Intel AboveBoard PS with 2MB of RAM, in addition to the 1MB of RAM on the motherboard. Cost, about $800.

E. For a VGA color graphics board and color monitor, add $750 to $1250. I recommend that you start off with a monochrome system, and upgrade to a color system after you become familiar with the computer, if at all.

F. You may want a printer. I recommend an inexpensive Epson LX-800 or equivalent, which presently costs about $200. After you become familiar with computing, you may wish to upgrade to a fancier printer, but at first this will be functional.

Both of these systems should come fully assembled. Ask if the hard disk is *FULLY ready to use*. If not, ask them to prepare it for you. (This preparation is called *formatting*.) Expect a small additional charge if this is not their usual procedure.

AT-compatible computers come with a *setup program* which you will have to run the first time you use the computer. This program creates a file that the computer reads each time it is turned on. This file tells the computer what equipment it has. For example, it needs to know if it has a hard disk, and if so, what type; it must know what the hard *drive specifier* is, whether it has a monochrome display, color display, or both, how many megabytes of RAM it has, etc. You can ask the mail order company if they can run the setup program for you. If they fully format the hard disk, put MSDOS on the disk, and run the setup program, you will have little else to do but plug the keyboard and monitor into the computer and turn it on.

The first-time buyer may expect to face a certain amount of anxiety. There are steps you can take to minimize this anxiety. The first step is to order your computer C.O.D. This way, you can at least be certain that no one has absconded with your money.

The real anxiety often arises *after* the computer is delivered. All mail order companies will provide both a manual and a customer support telephone number to call if you are having difficulties, and some give a 30-day, "no questions asked", money-back guarantee. Ask about this at the time of purchase. For those of you who are hesitant, there is another option available. Rent a friend knowledgeable about computers. Before buying the computer, look up "computer repair services" in your local Yellow Pages. Call one up, tell them that you are about to buy a computer mail order, and want to know what they will charge if you need assistance in getting the system going. Most computer repair businesses charge a flat hourly fee, with extra charges for parts. Since you will not need parts, you will be looking at the minimum hourly labor charge. If they charge $75 an hour, it will probably take them only a fraction of that time to help you format your hard disk (if it comes unformatted), or show you how to properly use some other piece of equipment. Even if it took two hours and cost $150, the money you have saved by buying mail order will pay the fee many times over. Besides, you will have the ear of an expert who can answer some of your questions.

Authorized Dealers

Happily, there is an even better solution. As I mentioned earlier, most of the digital photography equipment you will need must be bought from authorized dealers. In a photographic analogy, the difference between going to one of these dealers and going to a computer store is the same as going to a neighborhood one hour photo store for advice in solving a problem with your $2000 Hasselblad camera, or going to a professional Hasselblad repair service.

These authorized dealers are highly knowledgeable, and sell "high-end" digital photography and computer graphics equipment. If you are buying a digital photography color graphics board from them, they also sell the computer to put it in, along with all the other equipment you will need. Since you will probably have to use these people to buy your digital photography equipment sooner or later, you might as well have them set you up in the first place.

The only drawback to doing all your shopping with these specialized dealers is that they will probably charge you higher prices than standard mail order companies, although their prices may be lower than those at a computer store. One reason for the higher prices of specialized dealers is that they will probably sell you a "name brand" computer, instead of a *generic* computer. Another reason is that they will answer your questions as they arise. These companies are in it for the long haul, and want your continued business.

I bought my computer mail order and brought it to the specialized dealer who installed the specialized graphics board, and set up (installed) the specialized paint program, all at no additional charge. I got the best of two worlds: an inexpensive computer, with the specialized equipment professionally installed. Since I had no problems with the equipment they installed, I had no reason to contact them. About a month later, *they called me* just to check on how things were going. No sales pitch, just "We haven't heard from you, is everything all right?" While I can't guarantee that every dealer of specialized equipment is this concerned, it is nice to know that at least some are.

It may turn out that you do not live close enough to one of these specialized dealers to see them in person. If this is the case, you will have to deal with them by telephone. Unlike mail order companies, these businesses know exactly what type of computer you will need, and have probably put together numerous systems of the same type. You should receive a fully working system from these dealers.

If you are on a tight budget, try mail order. If you have more money available, and if you want to make the process as painless as possible, get to know these professionals who sell sophisticated graphics equipment, and take advantage of their expertise. Ways to contact these dealers can be found in the "Sources" section of this book.

In the following section we will discuss the various specialized equipment you will need, or may want to have, in order to make digital photographs.

PART II

HARDWARE AND IMAGE MANIPULATION

CHAPTER 4

GRAPHICS BOARDS (FILM)

The purpose of this chapter is to introduce you to specialized graphics boards suitable for use in digital photography. Just as the *base computer* described in Chapter 3 can be thought of as a camera body, so the graphics boards used for digital photography can be thought of as film. Just as your choice of film has a major impact on the final traditional photographic image, the graphics board you use will have a similar impact on your final digital photographic image.

This chapter is divided into four parts. First I will introduce a few new vocabulary terms, and expand on our discussion of *pixels* and *resolution*. Following this is a brief review of CGA, EGA, and VGA graphics boards used in standard computer applications. Next we will begin our discussion of specialized graphics boards. Much of this section is technical in nature, and for some it may be a bit taxing. Therefore, a number of short summaries will be given to help sort through the technical information to emphasize the major points you should focus on. Moreover, an attempt is made to limit the discussion to the practical aspects of these graphics boards: what they can do in helping to create an image, and not how they go about doing it. Finally, a brief summary of all the boards discussed will be presented, along with my suggestions of which boards you should consider, and why.

SOME NEW VOCABULARY

The concept of *high resolution* in digital photography entails two separate factors: the number of pixels and the number of colors that can be manipulated by the computer graphics board. Some graphics boards can control only a relatively few pixels and colors or only a relatively few pixels but many colors (low-resolution boards), some can handle a moderate number of pixels and a moderate number of colors (moderate resolution), and some can work with many pixels and many colors (high-resolution boards).

The number of pixels that appear on the computer monitor is determinec mainly by the computer graphics board, and to a lesser extent by the capabilities of the monitor itself. The graphics board can be thought of as a very complex set of switches that can turn on or off small lights on the computer monitor. These small lights form the image and are the pixels we discussed in Chapter 2. The design of the board determines the number of lights that can be turned on or off, and the number of different colors each light is capable of emitting. It should be obvious that the number of pixels the board is capable of manipulating will have an impact on the amount of detail that can be displayed on the computer monitor.

While the number of pixels a graphics board is capable of manipulating (turning on and off) is important, so too is the number of colors. If you draw a picture of a house on an 8 x 10 inch piece of white paper in black paint, the amount of detail will be less than if the same house were drawn with many colors.

In order to appreciate the differences among the graphics boards discussed in this chapter, it will be helpful if we discuss some new vocabulary terms. The pixels that do appear on the monitor screen are called displayable resolution (i.e., displayable pixels), since these are the pixels that are sent by the graphics board and actually shown (displayed) by the monitor. A number of the more sophisticated graphics boards are described not only in terms of displayable resolution (e.g., how many pixels they can send to a monitor), but also in addressable resolution (e.g., how many pixels they can store and manipulate). These boards are capable of containing an image larger than that which can be displayed on the computer's graphics monitor. Think of an image on an 8 x 10 inch piece of paper, on which is placed a 4 x 5 inch piece of clear glass (Figure 4.1). The image that comes through the glass, represented by the large rectangle, is equivalent to the *displayable resolution*, while the additional picture area would be equivalent to the *addressable resolution*, and would not be seen on the monitor.

It is also possible to have an image that is smaller than the amount of pixels displayed. This is similar to having a 4 x 5 inch negative and making a contact print onto a piece of 8 x 10 inch paper. The paper can contain more information than the negative contains. If you did have a smaller image than the graphics board/monitor displayed, the image would be surrounded by (usually) white pixels.

With many graphics boards, the addressable resolution and the displayable resolution are functionally the same. All the information contained in the graphics board is displayed on the monitor. With boards which have a greater addressable resolution than displayable resolution, the entire image cannot be seen on the monitor at one time. Instead, the image is viewed by panning, that is, using the computer to show different parts of the addressable resolution. Imagine sliding the clear glass around the larger image. You can see different parts of the image, depending on where you move the glass. The only restriction on panning is that you must have the edges of the glass parallel to the edges of the larger image. You cannot rotate the "window" you are looking through.

Although looking at the image in sections could take some getting used to, it is the way those graphics boards and systems capable of yielding the highest

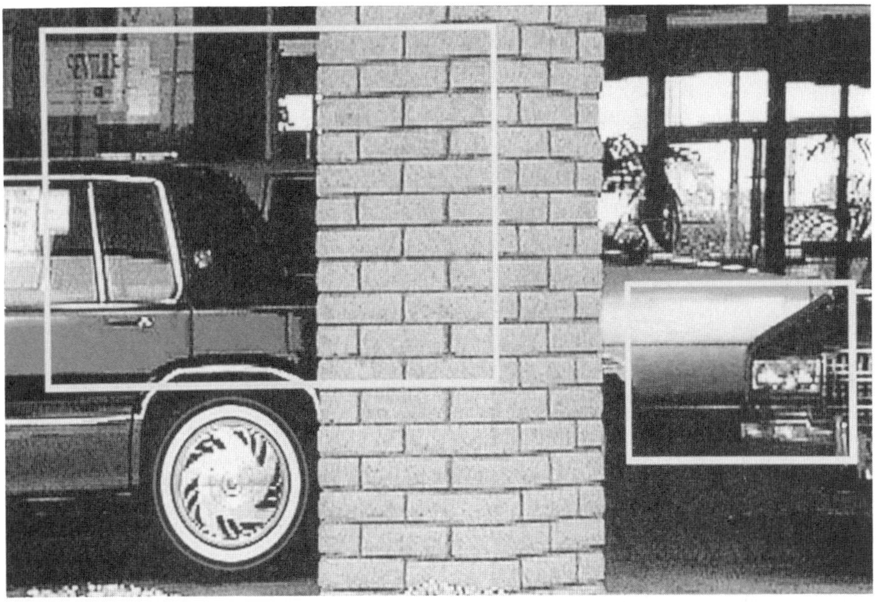

Figure 4.1 An image of a new car showroom shot with Canon RC-760 still video camera. Large rectangle represents about 1/4 of picture area, small rectangle represents about 1/16 of picture area. Image made with TIPS software.

resolution work. This is not so much a failing of the boards, but rather a limitation in the number of pixels current monitors are capable of displaying.

Besides the differences between displayable resolution and addressable resolution, there are also the differences between palette and palette. No, this is not a misprint, but a confusing situation which, I am convinced, is used by a small number of graphics board manufacturers to mislead potential buyers about their equipment.

The first meaning of *palette* is the total number of colors the graphics board is capable of making. The EGA class of standard (non-digital photography) graphics boards, for example, has a total palette of 64 colors.

The second meaning of palette is the number of colors the board is capable of displaying on the monitor in any one picture. This is sometimes called *simultaneously displayable colors*, or palette, and we will use this term to mean the number of different colors that you can have in any one picture. In standard computer use, the number of simultaneously displayable colors is often far less than the total palette. For example, the EGA boards that can create 64 colors can only display 16 colors in any one picture.

With most graphics boards used for digital photography, the total palette and simultaneously displayable palette are the same. Some of these boards are capable of simultaneously displaying either 32,768 (referred to as 32K) colors from a total palette of 32K colors, or simultaneously displaying 16,777,216 (16.7M) colors from a total palette of 16.7M colors. (In this book, I will use M for million.)

Some manufacturers, usually of standard graphics and monochrome boards used for everyday computing, will include in an ad for their product the statement, ". . . with a palette of 64 colors." You may assume that this is the number of simultaneously displayable colors. You might be right, but sometimes this statement actually refers to the total number of colors the board can manipulate (total palette), not the number it can simultaneously display at one time. Take nothing for granted. When purchasing a graphics board, whether for digital photography or for general computing, be certain to ask how many colors can be simultaneously displayed.

STANDARD GRAPHICS BOARDS

The Color Graphics Adaptor (CGA), as mentioned in Chapter 3, was the first color graphics system available for the IBM/PC line of computers. For these graphics boards and their clones, the addressable and displayable resolutions are the same, being 320 x 200 pixels. (Remember, 320 x 200 pixels means 320 pixels in each horizontal row, with 200 pixels in each vertical column.) This is very low resolution. At best, a photographic image looks blurry and lacks detail. The earliest use for this type of graphics board was for displaying business graphs and charts. While the quality of bar charts was acceptable, round "pie charts" appeared with "jaggies," which are jagged or saw-tooth-like diagonal or circular lines (Figure 4.2). *Jaggies* appear when there aren't enough pixels to create a smooth curve or

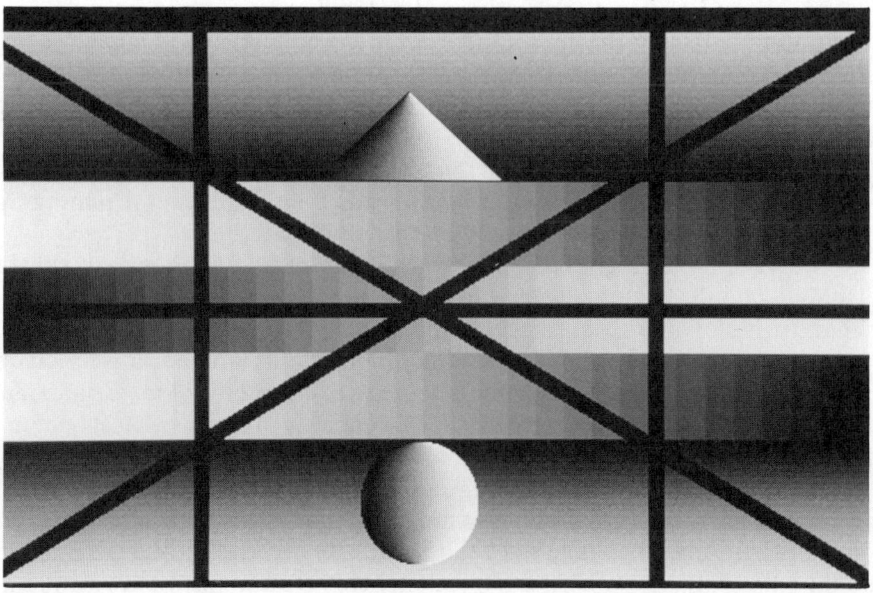

Figure 4.2 A focusing chart showing jaggies on diagonal lines and on edges of triangles and sphere, made with TIPS software.

diagonal. The total palette for a CGA board is 16 colors, with a maximum of four simultaneously displayed colors.

In 1984, three years after the introduction of the IBM/PC with CGA, EGA became the new graphics standard. It has an addressable and displayable resolution of 640 x 350 pixels, with a palette capable of simultaneously displaying 16 colors out of a total palette of 64 colors. Images generated by these boards also have jaggies, but photographic images made with EGA displays look less blurry and contain more detail then those generated by a CGA system.

In 1987, VGA was introduced, initially as the standard graphics board for IBM PS/2 models 50, 60, 70, and 80 computers, and as an upgrade for PS/2 models 25 and 30. IBM made available a VGA upgrade for older model IBM/ATs and within a few months, VGA was available for ATs from third-party manufacturers. VGA is usually described as having two modes, or two different resolutions. The first is 320 x 200 displayable and addressable pixels with up to 256 simultaneous colors out of a total palette of 262,144 colors. The second mode has 640 x 480 displayable and addressable pixels with up to 16 simultaneously displayable colors out of a total palette of 262,144 colors.

These two different modes are *software selectable*, that is, either a program is constructed to work in one of the two modes, or a program will allow the computer user to choose which mode to use. If you had such a board, you could choose between low resolution with many colors, or higher resolution with fewer colors. It should be noted that even at a resolution of 640 x 480 pixels, jaggies are still noticeable.

All these boards are capable of displaying text and graphic information on a computer monitor. With the development of *paint* programs, these boards were no longer limited to displaying business charts and graphs, but could also be used to create drawings. Although it is possible to use EGA and VGA boards in digital photography, their value is limited. First, the relatively low number of simultaneously displayable colors does not allow for continuous tonal changes in an image. Second, sophisticated programs for manipulating the images have not been developed for EGA and VGA boards. This is not to say that programs for manipulating images do not exist for EGA and VGA boards, but rather that a number of useful features found in more sophisticated programs designed for use with digital photography boards are not (yet) available for them.

Frame grabber boards take an image from a videotape, videodisk, or video camera producing an NTSC signal, *digitize* (it turns it into numbers), and store these numbers as a file in the computer. *NTSC* stands for *N*ational *T*elevision *S*tandards *C*ommittee, which is the standard signal for North American television transmissions. (PAL, which stands for *P*hase *A*lternation *L*ine, is the European standard for television transmission, and is incompatible with NTSC. All consumer grade TV cameras and accessories sold in the United States will meet the NTSC standard, while all consumer grade TV cameras and video accessories sold in Europe will be PAL standard.)

The digitizing of the TV image is accomplished by physically attaching the video camera, videotape, or videodisk player by a cable to a frame grabber-computer graphics board. You view the incoming TV images on the computer graphics monitor, just as if you were looking through the viewfinder of a camera. When the image of interest appears on the monitor, you press a key on the computer keyboard, or on a similar device, and the frame grabber board isolates (grabs) the signal, turns the colors and values (lightness/darkness) of the image into numbers, and stores the numbers in a file.

EGA/VGA Frame Grabber Boards

Within the past year or so, a number of frame grabbers have been marketed for use with EGA/VGA boards. Sometimes these frame grabbers are separate boards that do nothing other than grab an image, but increasingly this function is being built into standard EGA or VGA boards. The advantage of a frame grabber-graphics expansion board combination is the same as the advantage of any multifunction board: One board that can perform a number of functions frees expansion slots to be used with other computer boards. The combination frame grabber and graphics board can perform as any other graphics board of the same class (e.g., if it is an EGA + frame grabber, it can be used as a regular EGA board for standard computer use, or as a frame grabber board for use in digital photography). As previously mentioned, the disadvantage of EGA/VGA boards in digital photography is that they yield relatively low-resolution images (relatively few colors with relatively few pixels, resulting in unrealistic looking images with little detail), and many sophisticated programs which are designed to enhance the original image and help create quality photographic images are not available for them.

Most frame grabber boards are purchased directly from the manufacturer or from a mail order house, because they are not presently popular enough for most stores to keep in stock. The prime function of these EGA/VGA frame grabber boards is in desktop publishing, which includes in-house company newsletters and publications, club newsletters, etc. Images created for these purposes are usually inferior to newspaper-quality photographs. Most of these boards list for about $800, while similar quality EGA or VGA boards without a frame grabber function sell in the $250 range.

DIGITAL PHOTOGRAPHY BOARDS

Most graphics boards suitable for digital photography include frame grabbers. With one notable exception (discussed in the next paragraph), the boards mentioned below differ from EGA and VGA boards in that they have the ability to create high-resolution images.

The ICB (for *I*mage *C*apture *B*oard) is manufactured by *Truevision*, the leading manufacture of sophisticated digital photography boards. The ICB is a very low resolution board with a displayable and addressable resolution of only 256 x 240 pixels. However, it has a total palette of 32K colors, all of which can be displayed simultaneously. The incoming TV image can be adjusted for hue (color), saturation (rich or pastel colors), and intensity (lightness or darkness). Additionally, some special software (discussed in Chapter 7) is available to manipulate the captured image. The ICB does not function as a standard graphics or monochrome board for general computer use. The ICB lists for $1,295 and is available only through authorized Truevision resellers. Because of its extreme low resolution, the ICB is not recommended for digital photography.

TARGA Boards

Truevision also makes and sells a number of digital photography boards under the TARGA label. TARGA stands for *T*ruevision *A*dvanced *R*aster *G*raphics *A*dapter. There are a number of boards in the TARGA line and each has a special purpose. All are adequate for digital photography, and all share some common features.

All have a maximum displayable resolution of 512 x 482 pixels, although their addressable resolution is slightly higher (512 x 512 pixels). The slight discrepancy of displayable and addressable resolution is purely technical; you cannot *pan* a TARGA board to see the additional addressable pixels. This resolution, which at one time would have been considered high, is now considered medium. With these boards, the problem of jaggies still exists.

The boards are capable of 2x, 4x, or 8x *zooms*, or magnifications of the image. This is handy when you want to work on a small "delicate" area of an image, such as the hairs in an eyebrow. By zooming or magnifying a particular area, you can more accurately make modifications to the image. This is analogous to using a magnifying glass or "lupe" when retouching a negative or print.

Standard software programs (discussed at length in Chapter 7) used to manipulate the captured image are available for all but one of them, the TARGA-8.

TARGA boards do not function as standard graphics or monochrome boards.

TARGA boards are recommended to work with IBM AT computers and AT compatibles.

TARGA boards require two monitors, one for the image being manipulated (the digital photography long persistence monitor) and one to display instructions or information from the paint program (the standard graphics or monochrome monitor).

TARGA boards are capable of receiving video images.

TARGA boards can transmit video signals to other devices.

TARGA boards are sold only through authorized dealers.

Differences Among TARGA Boards

The *TARGA-8* is a black and white (B&W) frame grabber and image display board. It can simultaneously display 256 grays out of a total palette of 256 grays. Because it is a B&W board, it can be used with a less expensive gray-scale (B&W) monitor, instead of the more expensive color monitor. If you plan to work exclusively in B&W photography, this board might be worth considering, but it has at least one main deficit. There are no standard programs available to manipulate images created with this board. It accepts, and can send, an NTSC (North American Standard) but not PAL (European Standard) video signal. The TARGA-8 is mainly used for scientific purposes. The list price is $1,595.

The *TARGA-M8* is similar to the TARGA-8, although in addition to the 256 grays, it can simultaneously display 256 colors from a total palette of 16.7M colors. It too can be used with a gray-scale monitor, if you want to limit its use to B&W photography, but a more costly color monitor is needed for color images. While the 256 grays are more than sufficient for B&W photography, the 256 maximum simultaneously displayable colors will not permit a very realistic photographic image.

This board is capable of overlaying a live video image with a painted or drawn image. For example, a fantasy scene can be drawn on the monitor screen using a paint program, and a model could then be shot with a video camera and positioned (overlaid) accurately in the fantasy scene. Although the ability to overlay a digitized image with a live video image may sometimes be helpful, in practice this feature will rarely be used in digital photography. Its real intent is for use in television transmission, to allow for type (the name of a television show, for instance) to be combined with a live TV image.

Although there are a limited number of programs available for standard image manipulations with the TARGA-M8, not many manufacturers have designed software (programs) for it. This could limit your creative options. This board, too, is usually used for scientific purposes and comes only in an NTSC model. The list price of the TARGA-M8 is $1995.

The *TARGA-16* is the most popular of the TARGA series. It can simultaneously display a maximum of 32K colors from a total palette of 32K colors. This number of colors is adequate for full-color photographic images, and just barely adequate for B&W work. This is because the 32,768 colors are created from 32 values (lightness to darkness) of red, 32 values of green, and 32 values of blue: 32 x 32 x 32 = 32,768 different colors, but only 32 different grays. (To get a gray, you must have equal values of red, green, and blue. Since there are only 32 values of each color, and since the values must be the same for each color to create a gray, there are only 32 possible grays on a TARGA-16. If you have the same value of red and green, but a different value of blue, you won't have a gray, but a color.) A digital photographic image made up of 32 different grays ranging from black to white will look realistic under most circumstances, but not all. As was the case with the TARGA-M8, an *overlaid* live image can be merged with a painted or previously digitized photographic image.

The popularity of the TARGA-16 is due in part to its price. Its original list price was $2,995, but Truevision lowered it to $1,995 in an attempt to make it more affordable and increase the number of people engaged in digital photography and related activities. Because of its popularity, there are a relatively large number of third-party software programs available for it, some of which are very sophisticated. The creative tools for the TARGA-16 are almost unlimited. The TARGA-16 accepts both an NTSC and RGB (for *R*ed, *G*reen, *B*lue, discussed in the next chapter) video signal.

The *TARGA-32* is similar to the TARGA-16, but differs in a few important ways. First, it can simultaneously display a maximum of 16.7M colors from a total palette of 16.7M colors. This enlarged palette will give a *marginally* improved full-color image over the TARGA-16. It is difficult for many people to distinguish a full-color image made on a TARGA-16 from one made on a TARGA-32. It also has a marginally improved overlay capability, although again the difference is not dramatic. It will, in theory at least, noticeably improve B&W digital photography, since it will yield a 256 gray scale compared to the TARGA-16's 32 gray scale. Finally, because the TARGA-32 is not quite as popular as the TARGA-16, fewer third-party software programs are made for it, although such programs do exist, and more are being developed. The list price for the TARGA-32 is $4995. The TARGA-32 only accepts and sends an RGB video signal.

The *TARGA-24* is substantially the same as the TARGA-32, except that it does not have overlay capabilities. This means that you cannot overlay a live image with a previously digitized or painted image. There are relatively few third-party software programs available for the TARGA-24, limiting its creative potential, although some TARGA-32 programs will work with it. The list price for the TARGA-24 is $3995. As is the case with the TARGA-32, the TARGA-24 only accepts and sends an RGB video signal.

Summary of TARGA Boards

Boards of this type could be thought of as medium grain film (showing medium resolution). Those that are capable of displaying 256 simultaneous shades of gray (TARGA-8 and M8, 24, and 32) have excellent tonal ranges, and those capable of displaying 16.7M colors (TARGA-24 and 32) have excellent color characteristics. The TARGA-16 has an adequate tonal range and good to excellent color characteristics. Because of the exceptional number of third-party programs for the TARGA-16 and the smaller but increasing number of programs for the TARGA-32, these two boards are very well suited for use in digital photography.

ATVista and NuVista Boards

Truevision also makes higher resolution (more pixels and more colors) graphics boards for use in digital photography. The ATVista models work on IBM AT or AT-compatible computers, while the NuVista models work on the Apple Macintosh II computer. These graphics boards have a number of similarities, and share many of the common features of the TARGA boards listed above, but with a few differences.

They are all NTSC and PAL video compatible (unlike the TARGA boards, which are only NTSC or RGB video compatible).

They are all capable of displaying an image of 768 x 576 pixels, compared to the TARGA's 512 x 482.

Each of the Vista/NuVista boards is really three boards in one. This is because each has three possible resolution settings.

Additionally, they differ from each other, and from TARGA boards, in some important ways.

The ATVista 1M

The ATVista 1M (1M for 1 megabyte of video RAM) has the following three user-selected addressable resolutions.

1. The 1M can address 512 x 512 pixels with a maximum of 16.7M simultaneously displayed colors. This is functionally equivalent to the TARGA-24 and 32.

2. It can address 1024 x 512 pixels with a maximum of 32K simultaneously displayed colors. This is twice the addressable resolution found on the TARGA boards, and displays the same number of colors as the TARGA-16.

3. It can address 1024 x 1024 pixels with a maximum of 256 simultaneously displayed colors. This addressable resolution is four times greater than the TARGA series of computer graphics boards, but the maximum number of simultaneously displayed colors is severely limited.

Because the 1024 x 512 and 1024 x 1024 *modes* address more pixels than the 768 x 576 pixels the board is capable of displaying (sending to the monitor), the entire image is not viewable on the monitor at one time. Instead, *panning* is used to bring different sections of the image to the monitor screen. The 1M cannot be upgraded to perform like the 2M or 4M (discussed below). The list price for the 1M is $2995. Standard software is not available, nor are many third-party software programs useful for digital photography. The 1M will seriously limit your creative capabilities.

The ATVista 2M and NuVista 2M

The *ATVista 2M* (for 2 megabytes of video RAM) and *NuVista 2M*, like the 1M models, can operate in three different modes, and share the following characteristics.

1. They can address 1024 x 512 pixels with a maximum of 16.7M simultaneously displayable colors. This is twice the addressable resolution of the 1M in this mode.

2. They can address 1024 x 1024 pixels with a maximum of 32K simultaneously displayable colors. This is twice the addressable resolution of the 1M in this mode.

3. They can address 2048 x 1024 pixels with a maximum of 256 simultaneously displayable colors. This is twice the addressable resolution of the 1M in this mode.

The resolutions in items 2 and 3 can be considered high resolution, at least as of this writing. Because the maximum addressable resolution is so much greater than the displayable resolution of 768 x 576 pixels, you may be limited to viewing only a small portion of the entire image at any given time; this may cause problems when trying to view the overall composition of the image. As is the case with the 1M, there is a severe shortage of programs available for the ATVista 2M, at least as far as digital photography is concerned, although more are available for the NuVista 2M. The list price for both the ATVista 2M and NuVista 2M is $4250. The 2M series can be upgraded to a 4M for an additional $2000.

The ATVista 4M and NuVista 4M

The *ATVista 4M* (4MB of video RAM) and *NuVista 4M* can be set to operate at any of the following three modes.

1. Both of these graphics boards can address 1024 x 1024 pixels with a maximum of 16.7M simultaneously displayable colors.

2. Both can address 2048 x 1024 pixels with a maximum of 32K simultaneously displayable colors.

3. Both can address 2048 x 2048 pixels with a maximum of 256 simultaneously displayable colors.

Although the 4Ms do give high resolution, the inability to view the entire image remains a problem, which may cause serious composition problems. There is a fairly large third-party software market for the 4Ms. The list price of the 4Ms is $5995.

If you were in the market for one of the models in the ATVista and NuVista series, you would pay more than the amounts listed above. Truevision has announced surcharges due to an increase in the price of RAM chips. These surcharges are currently running from about 5% for the 1M to over 20% for the 4M, but this can change (either higher or lower) as the price of RAM chips fluctuates. The list price of the 4Ms with surcharge, for example, is currently $7495.

Summary of Vista Boards

As previously mentioned, each of these boards is really three boards in one. The user has the option of choosing exceptionally high resolution with limited colors (equivalent to extremely fine grained film with short tonal ranges), medium-high resolution with an acceptable number of colors (equivalent to a medium fine grain film with good tonal range), or medium resolution with a large number of colors (equivalent to a medium grain film with excellent tonal range). The main difference among the 1M, 2M, and 4M boards is the amount of video RAM included on the board available to manipulate images. The more *video* RAM the board uses, the more pixels the board can address (hold in its memory), and the more the board costs.

A Common Sense View of the Truevision Graphics Boards

For digital photography purposes, the main difference between the TARGA and Vista boards is the amount of addressable and displayable RAM the boards can manipulate. The ability of the Vista boards to address more pixels in their memory and to display more pixels on the monitor would ordinarily be an important consideration. If nothing else were to be considered, a general rule, "the more pixels and colors a board is capable of addressing, the better the quality of the resulting image" could be formulated. But as we will see in the coming chapters, there is a balance that must be struck between the size of a file and its ease of use. Some boards can address very high resolutions (e.g., manipulate files containing many pixels), but this does not necessarily make them more desirable than others that can address (manipulate) fewer pixels. In Chapter 7, for instance, we will discuss a number of different programs that have been designed to create high-resolution images from lower resolution images. With these programs it is possible to create images using TARGA boards with resolutions as high or even higher than those of images created with Vista/NuVista boards, and at the same time make the images easier to manipulate.

Thus the Vista/NuVista system does not necessarily offer images of higher quality than those available with the TARGA series. The money saved by buying a TARGA-16 can easily pay for the software to increase its resolution well beyond the capabilities of the Vista series. Furthermore, because of the large number of third-party programs suitable for digital photography that are available for the TARGA-16 and 32, these boards are more suitable for digital photography than the more expensive Vista boards.

It appears that the Vista boards are serving a function for those interested in animation: Many animation programs and much animation hardware are available for them. This is because the higher displayable resolution of these boards is about the same as found on a TV screen, unlike the TARGA boards, which have less pixels than TVs use. If animation is of interest to you, then consider a Vista board over a TARGA board, and in particular consider the Vista/NuVista 4M. However, if your interests are in still photography then I recommend that you consider a TARGA-16 "class" of digital photography board. The TARGA-32 is a *marginally* improved board, especially if you plan to specialize in B&W digital photography, but it may limit your creative options due to the slight decrease in third-party utility programs available for it. Since all the Truevision boards are sold by the same resellers, the merits of each in relation to your particular goals can be discussed in detail with the reseller.

You may have noticed a trend in the *horizontal* resolution of the TARGA and Vista boards. The TARGA boards all have a maximum horizontal resolution of 512 pixels per row, and the Vista 4M has resolutions of 1024 and 2048 pixels per row. All of these are binary numbers, that is, 2 taken to some power ($2^9 = 512$, $2^{10} = 1024$, and $2^{11} = 2048$). Digital photographers and computer graphic artists use the abbreviation 1K for a file with 1024 pixels per row, 2K for one with 2048, 4K for one of 4096, etc. As we will see in Chapter 6, under some circumstances an image may take up less than a complete row. For example, it is possible to have an image with 2000 pixels per row, instead of 2048. Regardless of the slight imprecision, resolutions of *about* 2048 pixels are referred to as 2K, and those of *about* 4096 are called 4K, etc. Resolutions of less than 1K are usually not abrevated, so a 512 resolution file is not called 1/2K.

TARGA-Compatible and Similar Types of Boards

Just as the original IBM PC/XT/ATs have become *cloned* by third-party manufacturers, the TARGA system has also been cloned. None of these clones are identical to the TARGA boards, and some are intentionally designed to offer features the TARGA boards don't offer.

Vision Technologies is a subsidiary of Everex, a well-known and respected manufacturer of IBM-compatible computers and computer hardware. Vision

Technologies offers three graphics boards that share many characteristics with the TARGA series, and one that is similar to the Vista series. Just like TARGA boards, Vision boards are sold through authorized dealers who will also sell you a computer and related equipment.

The *Vision-8* is similar to the TARGA-M8. It too can simultaneously display either a 256 gray-scale image or a 256 color image out of a total palette of 16.7M colors. It has a displayable resolution of 512 x 484 pixels, and contains a frame grabber for NTSC or PAL signals (but not both on the same board). The Vision-8A is similar to the TARGA-8, in that both can display only a 256 gray-scale image. Although software is available for the Vision-8 (but not the 8A) from Vision Technologies, third-party software support is limited, as is the case with the TARGA-M8 series. The Vision-8 currently lists for $895 for either the NTSC or PAL versions, and the Vision-8A currently lists for $850.

The *Vision-16* is similar to the TARGA-16. Like the TARGA-16, the Vision-16 can simultaneously display its total palette of 32K colors. Its displayable resolution is 512 x 484 pixels, it accepts either NTSC or PAL in addition to RGB video signals, and can also output these signals. It too can overlay a live video image with a previously digitized or painted image. A number of standard software programs designed for the TARGA-16 have been modified to run on the Vision-16. The Vision-16 has a current list price of $1,495, with a PAL version currently listing for $1,450.

The *Vision-32* is closer to the Vista series of graphics boards than to the TARGA series. It can simultaneously display its total palette of 16.7M colors. Its addressable resolution with 1MB of video RAM is 1024 x 512 pixels, and with 2MBs it can address either 1024 x 1024 or 2048 x 512. It can display either 512 x 484 or 646 x 484 pixels, depending on how you set the board to run. The Vision-32 is a two-board graphics system, which means that it occupies two expansion slots in your computer. The Vision-32 with NTSC or PAL currently lists for $2,695 with 1MB of video RAM, and $3,895 with 2MB of RAM.

The *Artist RealVision-16* from *Control Systems* can capture both NTSC and PAL signals, uses all popular software used by the TARGA-16, and displays 32K simultaneous colors from a total palette of 32K. It can address 512 x 512 pixels, and with a slight increase in its video RAM (which is available from the manufacturer as an option), it can address either 1024 x 512 pixels or 512 x 1024 pixels. The displayable resolution is 512 x 482 pixels with the standard amount of memory, but with the optional increase in video RAM it can display a varying amount of pixels, ranging from a high of 1024 x 482 to a low of 512 x 576, depending on how the board is set to run. The list price for the RealVision-16 is $3595. For the increased video RAM memory which will increase both the addressable and displayable resolution, add $895.

Control Systems also sells the *Artist RealVision-24*, which is similar to the TARGA-32. Its main difference from the RealVision-16 is that it can simultaneously display its total palette of 16.7M simultaneous colors. Both the Artist RealVision-16 and 24 can be optionally set up to also function as a standard EGA board, in addition to a digital photography board. The list price for the RealVision-24 is $4295, the memory upgrade lists for $1295, and the EGA module lists for $495 for both the 16 and 24 boards.

Making Sense Out of TARGA-Compatible Boards

I have already recommended that you consider the TARGA-16 (or 32) for your digital photography board. What do compatible boards have to offer? Should any of the boards I've discussed be seriously considered?

The Control Systems' Artist RealVision-16 and 24 boards do offer a feature that could be important to some readers. It is the ability to use CAD (*C*omputer *A*ssisted *D*esign) software. CAD is not usually thought of as a digital photography function. CAD is usually used to draw design plans for products as diverse as automobiles and furniture. The images may look like blueprints, or they may look like three-dimensional drawings of the finished products. The CAD images are created by the computer by taking instructions from the user for the placement of lines and curves; with some programs, even the placement of light sources can be specified in order to create highlights and shadows. Drawing skills on the part of the computer user are not necessary. You can literally be the type of person who can't draw a straight line with a straight edge and yet create "computer realistic" images with many CAD programs.

While there are available programs that allow the use of some CAD and CAD-like programs on a TARGA board, these CAD programs will work slowly. This is because a large number of mathematical computations are necessary to create a CAD drawing, and these computations are time consuming for the computer. If an occasional CAD drawing is to be done, the user can "suffer through" the sometimes very long time intervals needed to create the image on a TARGA board. But if CAD is to be used on a regular basis, then some means of speeding up the computations is necessary. For this purpose, specialized CAD graphics boards have been designed. The RealVision-16 and 24, although not full-fledged CAD boards, do have some CAD board characteristics that enable the RealVision-16 and 24 to run 600 times faster than the TARGA-16 and 32 boards. This means that the CAD drawings (as well as digital photography functions) will be performed more quickly. Moreover, the RealVision-16 and 24 are manufactured by a leading manufacturer of specialized graphics boards designed for CAD work. If you have no interest in incorporating CAD into digital photography, there really is no pressing reason to buy either the RealVision-16 or 24. However, if you decide that you want to merge computer-generated images with photographic images, then the increase in processing power (speed) will be beneficial, and a board such as the RealVision-16 or 24 could be a good buy.

The boards from Vision Technologies offer close compatibility with the TARGA and Vista series, with much lower prices. Because these models are still relatively new to the market (especially the Vision-32), less software is available for them than for the TARGA and Vista series. However, this is bound to change in the near future, and may have done so by the time you read this book. While the Vision-16 costs only about $500 less than the TARGA-16, which may not be enough to lure many people away from the more established TARGA-16, the Vision-32 offers a great reduction in price over the Vista 2M and 4M. (For comparison purposes, think of the Vista 32 as having abilities somewhere between the Vista 2M and 4M.) Moreover, for those who want or need a PAL board, the Vision boards should be considered, because the TARGA series is NTSC only.

SUMMARY

In this chapter we have discussed the pros and cons of various graphics boards for use in digital photography. In summary, the TARGA-16 and 32 are proven boards, with the TARGA-16 being the most popular of all the boards used for digital photography. The Vision-16's increasing popularity is indicated by the increasing number of programs designed to work with it. If you need an inexpensive PAL board, the Vision-16 should be your first choice, since the TARGA boards are NTSC only. If you are interested in combining computer drawings with photography, the Artist RealVision-16 or 24 should be considered. The Vista/NuVista has found a home with those interested in animation. For digital photography purposes, they offer no real advantage over the TARGA-16 and 32 or Vision-16 and 24 boards.

I realize that this introduction to digital photography graphics boards has contained information that may be confusing for some readers. For now, all that is important is that you realize that there is a wide selection of graphics boards that can be used for digital photography. Of all these boards, either the TARGA-16 or Vision-16 will be the graphics board most of you will buy. For this reason, I will concentrate on these two boards in the following chapters.

CHAPTER 5

VIDEO INPUT DEVICES (LENSES)

In the previous chapter we discussed graphics boards suitable for use in digital photography. These boards incorporated a *frame grabber* function which allows a video image to be digitized. In this chapter we will focus on video cameras as *input* devices for putting a photographic image into a computer. I will begin with a brief discussion of some new vocabulary specific to video. Then I will discuss consumer grade (home) video cameras, industrial-grade video cameras, one audiovisual device, and still video cameras. I will then briefly discuss how these cameras are used with the graphics board, and finally there will be a summary.

VIDEO VOCABULARY

Video cameras can be classified into at least three groups: consumer grade, industrial grade, and broadcast grade. Because broadcast grade cameras are beyond the budget of most individuals and moderate-sized businesses, we will limit our vocabulary to terms relevant to consumer and industrial grade cameras.

Consumer grade video (TV) cameras are the home video cameras we are familiar with. In North America, these cameras create a video signal that conforms to NTSC (*N*ational *T*elevision *S*tandards *C*ommittee) standards. In England and Europe, these cameras conform to the PAL standards for video transmission. Most consumer grade video cameras have a videotape recorder built into the camera, because NTSC (also referred to as *composite*) and PAL video signals can be recorded on videotape. Unfortunately, NTSC and PAL are incompatible systems. TV signals meeting NTSC standards cannot be used with PAL equipment, nor can PAL equipment be used with NTSC equipment. As we discussed in the previous chapter, some digital photography graphics boards can only be used with one of these standards, while others have been designed to accept signals from both.

Industrial grade video cameras usually use RGB (*R*ed, *G*reen, *B*lue) TV signals. RGB separates the red, green, and blue components of the scene being photographed, processes them separately, and finally recombines them to create a video image. Using a photographic analogy, NTSC or PAL is similar to making a color print by passing white light through a color negative onto photographic paper, while RGB is similar to making color separation negatives, and then printing each separation negative onto the paper. RGB yields a sharper image than either NTSC or PAL, especially at the edges of objects, although the RGB image usually does not contain more detail. RGB is incompatible with standard video recording equipment.

Because RGB images are usually considered to be of higher quality than NTSC or PAL, *decoders* can be used to convert an NTSC or PAL signal to RGB. It is questionable whether these conversions really add significantly to the quality of the image, at least as far as digital photography is concerned. *Encoders* can change an RGB signal to NTSC or PAL so that it can be used with standard tape recorders, or to convert an RGB video signal for use with graphics boards that can only accept an NTSC or PAL signal. Once the RGB signal is converted to NTSC/PAL, it loses the quality advantages of RGB.

Some encoders/decoders can convert between one or more signals. For example, if you use *Super VHS* home video equipment, a type of video signal that is not compatible with TARGA-type boards, an encoder/decoder could be used to take the Super VHS signal and convert it to a standard NTSC/PAL or RGB signal. Or, the NTSC signal from a North American video camera can be converted to a PAL signal to work with European video equipment. If you investigate an encoder/decoder, be sure to determine whether it will perform the conversions you need done, since some of these devices will not do all of the things I've just described. The cost of encoders/decoders starts at about $750.

Most consumer and industrial grade video cameras sold today are of the CCD type. CCD stands for *C*harged *C*oupled *D*evice, which in essence is a computer chip that contains numerous light receptors which react to the hue (color), saturation (richness of color), and luminance (contrast of the light) that strikes them. Technically, these receptors are called pixels, but I will refer to them as receptors to help avoid confusing them with the pixels that appear on the monitor's screen. While the pixels on the monitor are forming the image you see, the light that strikes the receptors on the CCD chip does not form the final video image. Information from the receptors is changed into a video signal, meeting either NTSC, PAL, or RGB standards.

The light receptors are arranged on the chip in a rectangular grid. The number of receptors in the grid helps determine the maximum detail the camera is capable of yielding. Many of the lower cost consumer grade TV cameras (usually those under $1,000) have only 250,000 receptors while some industrial grade and still video cameras may have 600,000 or more. The greater the number of receptors, the greater the potential detail of the resulting image.

Not only will the number of light receptors make a difference in the image output by the video camera, but so will the number of CCD chips. Consumer grade video cameras have only one CCD chip, while industrial grade video cameras may have from one to three CCDs. In two-chip cameras, one chip is used to detect the hue and saturation, and the other detects the luminance (contrast) of the scene. In three-chip cameras, different chips are used to determine the hue, saturation, and luminance. As the number of chips increases, so too does the quality of the image, and the price of the camera.

A major problem of one-chip and many two-chip cameras is a low contrast scale. These cameras tend to lose highlight and shadow detail, especially when the image is used by a graphics expansion board. The images have a good range of middle tones, but since they lack shadow and highlight detail, they tend to look "contrasty." Three-chip cameras give a better contrast range than one-chip and two-chip cameras, but even this range may be shorter than that of slide film, which is not known for having a long contrast range.

The sharpness of the video image is measured in *TV lines*. This is a rating system similar to measuring the sharpness of a camera lens. The more lines per millimeter a camera lens can distinguish when focused on a lens sharpness chart, the sharper the lens. The more TV lines a video camera can generate, the sharper the TV image. Most consumer grade video cameras can generate only 230 TV lines, although some are rated at 400 or more lines. Industrial grade video cameras range from a low of about 380 TV lines to 600 TV lines or more. Minimum broadcast quality (e.g., a sharp and detailed image) is usually considered to be 500 TV lines.

Therefore, the type of signal (NTSC/PAL or RGB), the number of light receptors, the number of chips used to form the image, and the number of TV lines generated all affect the quality of the image. Unfortunately, even if a video camera produces an excellent image on a TV set, the same image will be degraded when used with a TARGA-type board. This is because the TARGA-type boards generate only about one half the number of pixels of a TV set. The reduction in pixels causes a less detailed and less sharp image than the camera is capable of producing. In order to increase both the detail and sharpness of the image, software programs, described in Chapter 7, can be used. The effectiveness of these programs depends to a great extent on the quality of the original video image sent from the camera to the graphics board. The better the original video image, the more the programs have to work with.

If a Vista/NuVista-type board is used, the reduction of quality will be minimal, since these boards display the same number of pixels as a TV set. Although the Vista/NuVista-type boards will give superior images when compared to TARGA-type boards, I have not recommended them for a number of reasons. Even though the image will be sharper and more detailed, there is still the problem of a short contrast range. And if the contrast range of a particular three-chip RGB camera is satisfactory, the cost (usually $10,000 or more) will be prohibitive for many people. Even if the cost of a high-quality three-chip industrial grade camera and Vista/NuVista 4M board (probably a total price of about $18,000 or more) is no problem, the lack of available digital photography programs for these boards will be. In the next chapter we will discuss non-video methods for sending quality images to your digital photography graphics board for considerably less money than that spent for a Vista/NuVista and three-chip industrial grade video camera.

CONSUMER GRADE VIDEO CAMERAS

One advantage of consumer grade video cameras is their ability to put images onto videotape. This means that you can shoot on location, and then return to your computer and send the taped images into the graphics board. If you shoot ten minutes of videotape, you will have about 18,000 frames (images) to choose from.

Not all digital photography graphics boards can accept the NTSC or PAL signals consumer grade video cameras produce. The TARGA-24 and 32 need an RGB signal, so a decoder will be necessary to change the signals from a consumer grade NTSC or PAL camera to RGB for use with these boards.

Consumer grade cameras come in different formats, that is, they use different size videotape. Most consumer grade video cameras use 1/2 inch VHS videotape, but the smaller 8mm video cameras are becoming popular. Super VHS uses 1/2 inch tape, but puts out a signal that is not compatible with the standard VHS/8mm video signal. Both VHS and 8mm video can be easily input to a TARGA-16-type graphics board, but to use Super VHS for digital photography you will need to convert the non-standard signal to standard VHS or RGB with an encoder/decoder. For this reason, I don't recommend Super VHS cameras.

There are a number of differences among consumer grade cameras that you should be aware of. The most important is the ability to manually adjust the exposure. Many home video cameras are fully automatic, which is a sales feature to make the process of shooting easier for the average user. Although automatic exposure is handy, if you try to use the camera to copy photographs or slides into your computer, it will be very difficult, if not impossible, to get a usable image if the camera does not allow you to manually set the correct exposure. Frequently, the image sent to the computer will be either too overexposed or too underexposed to be of much use. The ability to manually adjust the exposure is essential.

A feature that seems to have become standard (at least with the higher cost consumer grade cameras) is multiple shutter speeds. A video camera shoots 30 frames per second, so it is not possible to shoot at shutter speeds slower than 1/30 of a second. Most of the better quality home video cameras have variable shutter speeds of 1/60, 1/100, 1/250, 1/500, and 1/1000 of a second or faster, although not all cameras have all of these speeds. Since higher shutter speeds will stop motion, thus providing a sharper image, you should make sure that the camera you select is capable of at least some of these higher speeds. Some lower priced consumer grade cameras only have shutter speeds of 1/60 of a second, which may be inadequate.

As of this writing, consumer grade cameras do not have interchangeable lenses. They all come with permanently installed zoom lenses, although auxiliary wide-angle and telephoto lenses may be purchased. These auxiliary lenses screw onto the front of the zoom lens. Most auxiliary lenses cost between $50 and $100, although some may go as high as $300.

You may be tempted to buy a particular camera because it is capable of making "posterizations" or other interesting visual effects. (A posterization is a black and white photograph that does not have a continuous gray scale, but rather a limited number of grays, or a color photograph with only a limited number of colors. See cover illustration.) Or you may see video accessories costing anywhere from $150 to $2,000, which will allow you to add visual effects and supposedly give you a sharper image. I wouldn't buy any of these devices, or cameras that incorporate them into their design, solely for these features. There are software programs (discussed in Chapter 7) that can run rings around even the best of these hardware goodies, usually at much lower costs.

Because consumer grade video cameras are single-chip cameras that have short contrast ranges, even the best may not produce digitized images that can be adequately enhanced by a software program. At the time of this writing, *Hi Band* consumer grade video cameras are coming onto the market. They use a wider range of frequencies to form the video image, which should help reduce problems of the short contrast range found in consumer grade cameras, at least to some extent.

INDUSTRIAL GRADE VIDEO CAMERAS

Industrial grade cameras are of higher quality than consumer grade cameras. They can create an NTSC or PAL signal, or NTSC or RGB signal, and sometimes both an NTSC (or PAL) and RGB signal. Most of these cameras do not have built in videotape recorder, because RGB is incompatible with present videotape technology. These cameras are useful for studio shooting of still lifes, or as copy cameras for digitizing prints or slides. They are not sold through standard "electronics" stores, or by mail order, but can be obtained either through specialized dealers or directly from the manufacturer. The prices of these cameras vary widely, from a low of about $1,300 to $20,000 or more.

The industrial grade cameras differ from consumer grade TV cameras in many respects. Most notable is the higher quality video signal, usually from 350 to over 600 TV lines, and the ability to create an RGB signal. Another major feature is the increase in the number of chips used to form the image. Although some industrial grade cameras are single chip cameras, many have two or three chips. The increase in the number of chips can help alleviate the problem of short contrast range found in consumer grade video cameras, but may not completely eliminate it.

Another advantage of industrial grade video cameras is that they allow the use of interchangeable lenses. Many of these cameras use standard "C" mount lenses, which will allow you to use 16mm motion picture camera lenses. There is a direct relationship between the size of the chip(s) used in a video camera and the size of an image on film. On 35mm film, the size of a negative or slide is 24mm x 36mm. Additionally, a 50mm focal length lens is generally considered to be normal on a 35mm camera, that is, it does not create either a wide-angle or telephoto image. A 2/3 inch CCD chip is about one quarter the size of a 35mm image. If you think of the size of the chip as the image size formed on film, a lens of 12.5mm focal length is needed to form an equivalent image to a 50mm lens on a 35mm camera. Since the chip is 1/4 the size of the image on film, a lens 1/4 the focal length of a lens on a 35mm camera is needed to give an equivalent perspective on the video camera as on the 35mm camera.

Although there are a vast number of industrial grade video cameras to choose from, JVC makes a number specifically designed to work with TARGA-type graphics boards. The JVC TK-870U is a single-chip RGB camera that can also output a standard NTSC signal. It has a standard "C" lens mount, which allows you to use 16mm motion picture lenses. This camera lists for about $1,400, without a lens. The JVC KY-15CI is a higher resolution 3-chip RGB camera. It can output RGB, NTSC, and Super VHS signals. The list price of about $6,400 includes a 10:1 macro zoom lens.

The JVC TK-F7100U is, by JVC's own admission, a rather strange video camera. It cannot be used as a standard video camera, because it takes 4/10 of a second to make an exposure. It does not use chips, but rather has a single tube. (Before the introduction of chips into video technology, video cameras used gas-filled glass tubes to convert the image coming in through the cameras lens into a video image. Today, many cameras are designed around computer chips, but some are still designed around tube technology.) The camera works by sending the red information to the computer graphics board, then the green information, and finally the blue information. It takes about 4/10 second for this process to be completed. Other RGB cameras send the red, green, and blue information in *parallel* (all at the same time), and hence can have very short exposure times. Due to the long exposure times, this camera is designed for making copies of prints, or for shooting still lifes. It lists for about $8,000, without a lens, and uses "C" mount lenses. Many Truevision and Vision Technologies resellers sell JVC cameras.

For those of you who are interested in B&W digital photography, it should be mentioned that B&W video cameras are available. Although they are not good for color photography, they do have two advantages over color industrial grade video cameras. First, they cost far less, usually from $500 to $1,500, and second, they usually have 500 TV lines of resolution or greater.

Of all the equipment mentioned in this book, industrial grade video cameras are the most variable as far as quality is concerned. Under no circumstances should you purchase such a camera based solely on technical specifications. Arrange to rent the camera you are interested in for a day, digitize some images, and see if you are satisfied with the results. If your Truevision or Vision Technologies reseller sells video cameras suitable for digital photography, get some digitized images that have been made with them, and view them on your own graphics monitor. Make sure the image quality is suitable for your purposes. Given the cost of this equipment, you should be certain that it will meet your needs before you buy it.

AUDIOVISUAL EQUIPMENT

A number of audiovisual devices which can be considered as industrial grade equipment have been developed to transfer slides directly to videotape. Devices such as these can also be used to transfer slides to the digital photography graphics board. The Fotovix II-X (pronounced "2X") is a slide transfer device made by Tamron. It consists of a base on which the slide or filmstrip is placed, and a zoom lens is located a short distance above it. To use the Fotovix II-X, you place a slide or filmstrip on the base; a fluorescent bulb within the base illuminates the slide; the image is picked up by the zoom lens, converted to an NTSC video signal, and sent to the computer via a cable. In addition to transmitting slides, it will also allow you to display color negatives as color positives. It has a control to help correct or create color shifts, and allows for up to a 6X zoom magnification of a section of the image. A large magnification may cause some graininess to appear in the image, but smaller image magnification will not cause graininess problems and can be used for cropping the image to isolate that part you are interested in.

The image signal is about 320 TV lines, which is better than the 230 TV lines of the consumer grade TV cameras, but not as good as industrial grade cameras. Additionally, because it is a single-chip device, digitized slides may appear contrasty, losing both highlight and shadow detail. The Fotovix II-X lists for about $1,800, it gives a much improved image over consumer grade TV signals, and is a slick looking piece of equipment (which may be of importance to those who need to rapidly digitize images in the presence of clients). In time, devices such as this will improve in image quality, to the point where they may become standard equipment for use in digital photography.

STILL VIDEO CAMERAS

Still video cameras are in their infancy, but already offer the digital photographer the means to create and put into a computer images suitable for digital photography. As the name implies, these cameras are a cross between traditional still cameras and video cameras. Instead of shooting "motion video," they are designed for still photography. Amateur still video cameras usually cost under $1,000. Because the quality of their images is not suitable for our digital photography purposes, we will concentrate on professional still video cameras. At present, there are at least two such color cameras that are readily available, with a third scheduled for sale by the time this book is published.

One is the Canon Still Video Camera RC-760, which costs about $5,500. It looks like a flat, (wider than it is deep) motorized 2 1/4 inch single lens reflex camera, with an enlarged right side which serves both as a pistol grip and a housing for two NiCad batteries. Two fully charged batteries are said to be capable of 400 exposures. It has a fixed eye-level viewfinder, although some right angle adaptors can be attached to the back of the eyepiece. The viewfinder shows a digital readout of exposure information. The RC-760 has 600,000 light receptors on a single chip, which gives very good detail when compared to most of the cameras mentioned above. The NTSC signal is about 360 TV lines, which is fairly sharp, and can be further sharpened with computer software.

The camera has shutter speeds ranging from 1/8–1/2000 second, is synchronized for electronic flash at 1/80 second, shoots either single exposures or 2, 5, or

10 frames per second, automatically color corrects (called *white balance*) for indoor/outdoor shooting, has a center-weighted TTL (*Th*rough *T*he *L*ens) light meter, allows for shutter-priority automatic exposure, and has the equivalent ISOs (light sensitivities) of either 80, 160, 240, or 480. The camera has a bayonet lens mount and can accept standard 35mm Canon FD and Nikon lenses with optional adaptors.

The light receptors are on a single 2/3 inch CCD chip, which means that the "normal" focal length lens would be 12.5mm. This is equivalent to a 50mm lens on a 35mm camera. Because a 12.5mm lens is normal, it becomes difficult to use the camera for wide-angle shots, because lenses with focal lengths of less than 12.5mm are very rare. In 35mm photography, a 12.5mm lens is considered an extreme wide-angle lens. Canon makes a 6mm f/1.8 lens for the RC-760 which is equivalent to a 24mm lens on a 35mm camera. This lens lists for $1,129. Canon also makes an 11–66mm zoom (equivalent to a 44–264mm on a 35mm camera) for $899, and a 50–150mm zoom (200–600mm on a 35mm camera) for $999.

The image created by the camera is put onto a 2 inch floppy disk. Think of these disks as flat videotape, instead of videotape rolls. The disk is inserted into the camera similar to the way cut film is inserted into the back of a view camera: Just slide the disk in place and start shooting. The disks list for about $10 each, and are usually sold in packages of ten. As with videotape, the information on these disks can be erased and the disks reused. It may be possible to use a disk 100 times before image degradation becomes noticeable.

The RC-760 is capable of shooting in two different modes. The first is *frame* and the second is *field*. A video image, whether sent directly from a camera or from a videotape/still video player, is put on the screen in two sections. First, the odd lines that make up a video image are displayed, and then the even. It takes 1/60 of a second to display one of these sections, called a field, and 1/30 for both fields to be displayed (one frame). Obviously, a field contains only half the detail of a frame. The RC-760 will permit you to shoot in either field or frame mode. If you choose field, you will not get the quality of a full frame, but you will get 50 exposures on a disk instead of the 25 in frame mode.

The camera is not capable of sending an image directly to the computer graphics board, and so a recorder-player or player is needed. The RR-450 digital image recorder-player (list price about $1,360) reads the image on the disk and sends it to a conventional TV or to the computer. The recorder-player has an NTSC signal of 360 TV lines. While the RR-450 will both record TV and video images and play back images made with the RC-760, the less expensive RV-301 player (list about $900) just sends images made with the camera to the computer or TV.

Another still video camera is Sony's *ProMavica* camera, which Sony markets through authorized Truevision (TARGA/Vista) dealers. As a 2/3 inch single-chip NTSC still video camera, it shares a number of features with the Canon RC-760. It uses the same standard 2 inch disks as the Canon RC-760, and allows shooting in field or frame mode. Its range of shutter speeds is similar to the RC-760, being 1/15 to 1/1000 of a second, with flash synchronization at 1/250 second. The ProMavica fires in single exposure mode, at 4 frames per second, or continuously. The ProMavica looks similar to a conventional 2 1/4 inch single-lens reflex, since it is deeper than it is wide.

The ProMavica differs from the RC-760 in some important ways. First, it has a non-removable 12–72mm f/1.4–1.7 zoom lens (equivalent to a 48–288mm zoom on a 35mm camera), it has only 380,000 light receptors compared to the RC-760's 600,000, and has a resolution of 320 TV lines instead of the 360 TV lines of the Canon. Because it uses a 2/3 inch CCD chip, the effective focal length of a 35mm camera lens is increased four times. This means that lens interchangeability may not be an important feature. An 8mm fisheye (extreme wide-angle) lens for a 35mm camera has an effective focal length of 32mm (semi-wide-angle) lens on a 2/3 inch chip camera. There are few lenses made for 35mm cameras with focal lengths shorter than 8mm. Therefore, lens interchangeability will not allow you to create extreme wide- angle images. The 12–72mm (48–288mm 35mm camera equivalent) non-removable lens will probably be suitable for most photographic needs, and supplemental (screw-on) lenses can be used to increase or decrease the focal length of the ProMavica's lens.

The list price of the ProMavica is $3,995, or about $1,000 less than for the Canon RC-760, but the recorder-players that are available to transfer the video image to a computer graphics board list from $2,680 to $3,940, which is two to three times more expensive than the equivalent equipment from Canon. Of these two cameras, the Canon RC-760 is less expensive (for both the camera and the needed player), and gives a higher quality image than the Sony ProMavica (600,000 light receptors and 360 TV lines compared to 380,000 light receptors and 320 TV lines). Because still video cameras and equipment are fairly standardized, it is possible to purchase a Sony ProMavica for about $4,000, and a Canon RV-301 player for about $900, thus reducing the overall cost of a still video system to under $5,000.

As for general quality, these still video cameras are considerably sharper than the 230 TV lines of resolution of most standard consumer grade video cameras, and offer about the same resolution as some of the industrial grade video cameras. While their images are not "tack sharp", they are in the usable range. But, as is the case with all single-chip cameras, the resulting image from still video cameras may lack both shadow and highlight detail, especially if the lighting is contrasty. It is best to use these cameras with very flat light.

Finally, Sony has announced a two-chip *Hi Band* color still video camera with a 500 TV line signal, which should be available for sale before the time this book is published. Like the Canon RC-760, it too uses interchangeable lenses. Its price without a lens will be about $6,500, and a zoom lens will cost about $3,500. While a standard still video player can be used, a Hi Band player (list price about $2,800) will be needed to get the extra sharpness the camera is capable of delivering.

With the exception of their current lack of high resolution and their short contrast range, the Canon and Sony professional still video cameras are in every other sense traditional cameras. They have electronic flash sync, a full range of traditional shutter speeds, through-the-lens light metering, and are capable of rapid sequence shooting. In short, shooting with either camera will be substantially the same as shooting with a traditional hand-held camera.

INPUTTING VIDEO IMAGES

Inputting an image from a video camera to a digital photography board is a simple matter. If the camera is NTSC or PAL, a standard video cable is attached to the *video out* plug on the video camera and connected to the *composite video in* receptor on the video graphics board. If you are using an RGB video camera, a special RGB cable is connected to the camera and to the *RGB in* plug on the graphics board. If a still video camera is used, the video disk is placed into the video player which is connected to the *composite video in* plug of the graphics board. Using a program (discussed in Chapter 7), you view the image being sent from the video camera or player on the computer's graphics monitor. When the image you want to grab appears, you press a key on the computer keyboard (or on a similar device) and the image is captured in 1/30 of a second. You can also digitize images directly from a television set. To do this, you can either plug one end of a cable into the TV (if it is one of the newer models that has a *video out* plug) and the other into the computer graphics board, or connect the computer graphics board to the *video out* plug of a videotape recorder. In this mode, you are using the live image coming into the tape recorder (the same way you use the image to tape a program), but you are not taping the image, just transferring it to the computer.

SUMMARY

Using a video camera is the easiest, and for the photographer, the most logical way to input an image into a computer. Unfortunately, at present there are two main problems associated with this method. First, some of you may find that the image is neither sharp nor detailed enough, and second, with almost all affordable video equipment, extremely low-contrast lighting is required to avoid losing highlight and shadow detail. Undoubtedly, these problems will be overcome as video technology advances, but the design of TARGA-type boards also contributes to the lack of sharpness and detail and the low contrast ranges typical of video digitized images. If you should see a demonstration of a video camera's image on a TV set, remember that the same image will be degraded when digitized with a TARGA-type board. If video cameras are to be used as an input device, the only alternative at this time to maximize image quality is to use a Vista/NuVista board at about four times the cost of a TARGA board, and do without a large base of programs available to manipulate digital photographs. Moreover, those programs that are available for the Vista/NuVista boards tend to cost 50% to 200% more than the same program for a TARGA-16 board. For these reasons, I have not suggested you get a Vista/NuVista graphics board.

In the next chapter, we will discuss a very different method to digitize photographic images, which produces extremely high-resolution images on TARGA-type boards.

CHAPTER 6

SCANNING INPUT DEVICES (LENSES)

In the last chapter we discussed using video equipment to put a photographic image into a digital photography graphics board. In this chapter I will discuss a method of inputting an image that involves the use of scanners. First I will discuss what scanners are, how they work, and how you would use one; in this section I will also explain the relationship between photographic film resolution and the resolution created by scanners. Then I will explore gray-scale scanners, and color scanners. I will then talk about some problems associated with scanners and digital photography. I will discuss the role of service bureaus, which are the equivalent of photo labs for digital photographers. Finally, there will be a brief summary.

AN OVERVIEW OF SCANNERS

Scanners comprise a class of computer hardware that can create extremely high resolution digitized images from B&W or color prints, negatives, or slides. Many scanners look like desktop photocopy machines. A photograph you want to digitize is placed face down on the glass platen of the scanner, in the same way you would place a piece of paper on a photocopy machine. A long CCD chip with a single row of light receptors moves underneath (*scans*) the photograph, changing the shades of gray or colors in the print into numbers. These numbers are sent via a cable to the computer, where they are stored in a file. This file can be used by software (discussed in the next chapter) to manipulate the numbers, and thus change the image.

In order to operate the scanner, you will use a program, technically called a *driver*, which either comes with the scanner or may have to be purchased separately. The driver acts as a translator between the scanner and the paint program used to manipulate the digitized file. Most paint programs need information (the numbers created by the scanner) to be arranged in a specific form. Because there are so many different types of scanners, and so many different types of paint programs, incompatibility problems arise. Different drivers are designed to take information from a particular type of scanner and translate it into a form a particular type of paint program can understand. This is similar to having a camera with a lens that is designed to fit a different camera. If you want to use the incompatible lens and camera, you will need an adaptor to connect them together. Drivers perform a similar function.

When you want to scan (digitize) an image, you type the name of the driver you use to operate the scanner on the computer keyboard. Usually, a menu will appear on your standard monitor's screen. It will display a number of options—often, but not always, in *icon* form. Icons are cartoon-like drawings, made popular by the first Apple computer. These options are used to tell the driver how you want the image digitized. Think of these options as shutter speed and f/stop settings on a camera. Different settings will create different looking images. Explanations of the different settings and examples of standard settings are given in the manual which is included with the driver and/or scanner. After you have answered the questions on the menu, the driver will operate the scanner and create the proper file for use with your digital photography paint program.

The scanner sends the information detected by the strip of the light receptors to the computer via a cable. The cable plugs either into the serial or parallel I/O port, or into a special expansion board which comes with the scanner. A number of scanners use a GPIB (*G*eneral *P*urpose *I*nterface *B*us, also called an IEEE-488) expansion board. This board must be bought separately from the scanner, and lists for about $750. GPIBs are specialized I/O boards for use with sophisticated graphics and scientific equipment. One board can be used with numerous different types of graphics input and output devices. If you buy a GPIB for a scanner, and later buy another device that also needs a GPIB to operate, you won't have to buy another GPIB, since one can be used with many different devices. GPIBs are sold by the same people who also sell the scanners that need the GPIBs to operate.

Although there are many different types of scanners, only a few types of scanners are of value in digital photography. For example, some scanners are designed to read typed, type set, or even handwritten information. These scanners are called OCRs, which stands for *O*ptical *C*haracter *R*ecognition. OCRs are used by businesses to feed typed information into their computers, thus saving the time it would take to retype it. OCRs are of no value in digital photography.

Graphics scanners are designed to put B&W or color photographs, paintings, or drawings into a computer. These are the type of scanners useful in digital photography. Combination OCR and graphics scanners are capable of acting either as an OCR or graphics scanner. Generally speaking, a graphics-only scanner is best for digital photography, since it can be argued that a scanner that tries to do two different things might not be able to do either of them as well as a scanner designed to do only one thing.

As mentioned in Chapter 2, the image on a computer monitor is created by a *rectangular grid* of pixels, which is wider than it is high. In Chapter 4 the rectangular grid (*displayable resolution)* of the different digital photography graphics boards was listed. TARGA boards, for example, display a maximum of 512 pixels per horizontal row x 482 pixels per vertical column. Scanners designed to digitize prints allow you to vary the number of pixels they will use to create the image. You can usually set the scanner to create anywhere from 30 to 300 dots per inch (dpi). *Dots per inch* is a desktop publishing term, and will make more sense to us if we think of them as *pixels* per inch.

One reason for varying the number of pixels per scanned image is to allow you to create a grid from your original photograph to match the displayable resolution of your digital photography graphics board. For example, if you have made a full-frame print from a 35mm negative on 8 x 10 inch paper, the resulting image would be 10 inches wide by 6.67 inches high. You could tell the driver to scan the print at a resolution of 51 pixels per inch, for 10 horizontal inches and for 6.67 vertical inches. This would create a file containing 510 pixels per horizontal row x 340 pixels per vertical column. The 510 horizontal pixels are only two short of the maximum displayable resolution of 512 pixels. The 340 vertical pixels per column are 142 pixels less than the 482 vertical pixels the TARGA board can display. This doesn't matter, since the proportions of the full-frame 35mm print are different from the proportions of the video monitor. What happens is that your image will have some "empty" space surrounding it on the top and bottom (and one empty space on either side of the image), but the entire 10 x 6.67 inch image on the print will appear on the monitor. We will discuss ratios of film and video monitors in more detail later in this book. For now, just remember that the variable amount of pixels per inch setting of the scanner will enable you to match the size of a print to the displayable resolution of your monitor.

Having adjustable pixels per inch might not be important if all prints were a standard size, but they aren't. If you had made a 5 x 3.34 inch print instead of the 10 x 6.67 inch print, a setting of 51 pixels per inch would create an image of 255 pixels per horizontal row x 170 pixels per vertical column, which would create an image only about 1/6 the size of your monitor's screen. To enlarge this smaller print (and thus give you more picture detail to work with), you would set the scanner, through the driver, to digitize at 102 pixels per inch. This would result in a file of 510 pixels per horizontal row x 341 pixels per vertical column.

A Note About Film Resolution Versus Digital Photography Resolution

There is another, very different reason for having adjustable pixels per inch. This involves the concept of matching the resolution of the film the image was made with to the resolution of the digitized image. While a digitized image can be defined by the number of pixels contained in the grid that makes up the image (e.g., 512 x 342), film does not have pixels, and so a direct comparison between the two systems becomes difficult.

The question is, how many pixels would film have, if film had pixels? The answer to this question is that there is no *single* answer. Film is rated in *resolving power* (i.e., how many lines per millimeter it can distinguish), but the actual resolution (sharpness and detail) of a photographic image is a product of the resolving power of the film *and* the resolving power of the lens used to make the image. (Additionally, other factors, such as the resolving power of the lens at the actual f/stop used to shoot the picture, which may vary greatly from the rated resolving power of the lens specified by the manufacturer, and the focus of the lens and the proper or improper development of the film, all have an effect on the actual resolution of the final negative or slide.) Thus, different films and different lenses create different quality images. No single number can represent the amount of pixels film would have if film had pixels. Regardless, I have been told by experts in the digital photography field that 35mm film has the equivalent of a total of 3.5 million pixels. My own calculations, based on the specifications for Kodak 35mm Varicolor film and a Nikon 50mm autofocus lens, show that there would be between 2.5 million and 3 million pixels in the negative, depending on the f/stop the lens was used at. A different lens would yield different results.

Let's say you made a 10 x 6.67 inch print from a 35mm negative. If you scanned the image at 51 pixels per inch, you would have a grid of 510 x 340 pixels.

The grid of 510 x 340 pixels = 173,400 pixels. This is considerably less than the 2.5 million to 3 million pixels my film and lens combination yielded. The image will still be good (due to the excellent quality of images produced by scanners), and there are software methods, described in the next chapter, that can be used to increase the number of pixels in the grid to match the resolution of the original negative.

However, since the print scanner has an adjustable scanning rate of 30 to 300 (or more) pixels per inch, you could set the scanner to read 204 pixels per inch. At this resolution, you would get 10 inches x 204 = 2040 pixels per horizontal row, and 6.67 inches x 204 = 1360 pixels per vertical column. This grid of 2040 x 1360, or 2,774,400 total pixels, falls within the 2.5 million to 3 million pixel range calculated for the film and lens combination mentioned above. Thus, the file containing the scanned image would have about the same resolution as that film and lens combination. We will now define high resolution as a file that contains about 2.5 million or more pixels. Thus, files in 35mm ratio of 2K (2048 horizontal pixels) resolution or higher are considered high resolution.

The file containing information for 2.7 million pixels is too large to be displayed by a TARGA board, which can only display a maximum of 512 x 482 pixels, and that would display only 510 x 340 pixels for our 35mm example. It would seem self-defeating to create a high-resolution file of 2040 x 1360 pixels that couldn't be used by the TARGA board. However, some software methods, described later in this chapter and in Chapter 7, will allow large files to be used with TARGA and TARGA-type boards.

We will have much more to say about file sizes and the proper ratios of film to the video monitor throughout this book. For now, keep in mind that variable pixels per inch scanning serves two purposes. First, it allows different size prints to be matched to the maximum displayable resolution of your digital photography board, and second, it allows you to create high-resolution files, which can also be displayed by the board with the proper equipment.

You may be wondering why the size and ratio of the image was of no concern when we discussed video. This is because the TARGA-type boards are modified television sets, and the Vista/NuVista boards display almost the same amount of pixels in the same ratio as television sets. The video signal is automatically "fit" into the space provided by the digital photography board, in a similar way a television signal is automatically fit to the size of a particular television set.

Gray-Scale Scanners

Gray-scale scanners will produce a file containing black and white information from a color or B&W print. Scanners of this type are of value if you want to limit your work to B&W, or if you want to start with a B&W photographic image and "hand color" it. Most gray-scale scanners are not designed for use with TARGA and TARGA-compatible boards, but are designed for use in desktop publishing. Desktop publishing involves taking the digitized image and either making it into a print by using an office gray-scale laser printer, or using a Linotype or similar typesetting/graphics machine to make printing plates for offset lithography. Although most gray-scale scanners are not designed for digital photography, utility programs have been written to convert a file created by a gray-scale scanner, which is incompatible with TARGA-type files, into a TARGA-type file. These utility programs are not drivers, since they do not operate the scanner. They are just conversion programs that transform the file created by the scanner into a different form which is meaningful to the paint program used to manipulate the image. These utilities usually cost between $100 and $200. Additionally, some manufacturers of gray-scale scanners offer their own conversion programs.

Gray-scale scanners differ from each other in at least three ways.

The first is that line scanners create high-contrast images only. If a particular gray or color is lighter than a particular setting set on the scanner (for instance, above a middle-tone gray), it becomes white, and if the gray or color is equal to or darker than a middle-tone gray, it becomes black. This is similar to using a high-contrast film such as Kodalith to make a high-contrast image from a B&W or color negative or slide. Scanners of this type are fine for graphics line art work (high-contrast images), but for full-color or full-gray-scale work, they are inadequate.

Secondly, those gray-scale scanners that are not limited to only B&W differ in the number of grays they are capable of recognizing. Some of these scanners can only reproduce a maximum of 16 grays. If you try to scan a full-scale B&W or color photograph with such a scanner, it will assign a particular gray tone or value of a color in the original image to the closest gray it is capable of creating. Full tonal range images scanned with a 16-level gray-scale scanner will appear slightly posterized. Other gray-scale scanners can digitize 32, 64, and sometimes 256 distinct grays. A scanner that can reproduce a larger number of grays will be able to produce a digitized image that is very similar to the original image. For the best results, a 64 or 256 gray-scale scanner is needed.

Thirdly, not all "gray scales" are the same. Some gray-scale scanners do not reproduce a true gray scale, but rather a *dithered* pattern. This is a method of positioning black dots and white dots in such a way as to try to fool the eye into interpreting the pattern as various shades of gray. For example, two black dots surrounded by six white dots will appear as a very light gray, while one white dot surrounded by seven black dots will appear as a very dark gray. Different scanner manufacturers use different dithering schemes to create the optical illusion. In actuality, *any* dithering pattern is not as good as a true gray scale image. If you should decide to purchase a gray-scale scanner, be sure to find out if the image it produces is a true gray-scale, or a dithered pattern. If the latter, I suggest you look for another scanner.

Since desktop publishing is a large, well-established field, *many* gray-scale scanners are made, so there is a wide range from which to choose. The Hewlett-Packard ScanJet Plus may be the most popular. It can create a 256 gray-scale image from a print up to 8.5 x 11.7 inches, at 12–1500 pixels per inch resolution. It can enlarge sections of the image up to 200%, and it will work with color prints (although it doesn't work well with some yellows, so it's best to use a B&W original). The ScanJet Plus lists for about $1,600. An interface kit, including an expansion board and driver, must be bought from Hewlett-Packard to make the ScanJet Plus work. Depending on the interface kit purchased, the ScanJet Plus will work either with the IBM XT/AT/or PS/2 or Macintosh Plus/SE/II models. The kit lists for about $600. Both the ScanJet Plus and interface kit can be bought discounted by mail order (currently about $1,500, including the needed interface kit). The output file from the driver is not compatible with TARGA-type boards, so a utility conversion program, such as *SCAN-T* by Videotex Systems ($149), will be needed to convert the file created by the ScanJet Plus to a TARGA-16 format.

Since there are so many other gray-scale scanners available, it's best for you to ask your TARGA or Vision Technologies reseller for their recommendations as to which ones may be best for your purposes.

Besides full-sized gray-scale scanners, a number of *hand-held* scanners have been introduced to the market over the past few years. To operate these scanners you roll the scanner over the photograph you want to digitize. Some of these scanners will only digitize a maximum area of 2 1/4 x 10 inches, while others will create a maximum 4 x 6 inch digitized image. All of these scanners use dithered patterns, (e.g., they do not give a true gray scale), and the images they create are of relatively low resolution. These scanners are relatively inexpensive, usually in the $200 range, including the software and cable needed to run them. As is the case with the ScanJet Plus and most other B&W scanners, a utility conversion program will be needed to change the file these scanners create to a TARGA or TARGA-compatible file. These scanners are not recommended for quality digital photography, but they might be of interest to artists, or for purposes of experimentation. (Rolling them randomly over a photograph may create some interesting visuals, for example.)

COLOR SCANNERS

Color scanners are similar to gray-scale scanners; the major difference is that they are capable of determining not only the value of a small section of an image, but also its color. Color scanners tend to be specialized to digitize either prints or slides and negatives.

While there are many gray-scale print scanners to choose from, you choices will be limited for color print scanners, although more companies are beginning to enter this field. The Howtek Scanmaster (list price about $7,000) is capable of creating TARGA files with 16.7M simultaneous colors, from prints and other flat art. It can scan prints up to 11 x 17 inches in size, and can also scan slides and transparencies up to 8 x 10 inches with an optional attachment (list about $650). The number of dots per inch ranges from a low of 30 to a high of 300. Howtek also sells the Scanmaster II (list about $5,000), which is similar to the Scanmaster but can only scan a maximum area of 8.5 x 11 inches. A GPIB must be purchased for either of these scanners, raising the cost by about $750. A full-color print can be digitized by either of these scanners in less than 5 minutes.

The Sharp JX-450 color scanner (list price about $7,000) is substantially the same as the Howtek Scanmaster, and the Sharp JX 300 (list about $5,000) is substantially the same as the Scanmaster II, since both Scanmasters are slightly modified Sharp color scanners. This is not a particularly unusual situation. Major manufacturers, such as Sharp, sometimes enter into agreements to sell some of their products to other companies, which then may make cosmetic changes to the product and sell it under their own label. Howtek, unlike Sharp, advertises heavily in computer graphics magazines, and is the scanner that has a wider name recognition among computer graphics users.

The Scanmasters and Sharps are relatively simple to use. You place the print on the glass platen in the same way you would place a print on a photocopy machine. You then call up the driver used to control the scanner. Howtek sells one driver called Scan-It (list price about $1,000), which uses icon menus. With this driver you can control the size of the area of the print to be scanned by having the driver display the image to be scanned on the digital photography graphics monitor. Using the computer keyboard or other device (discussed in Chapter 7), you can draw a box around the area of the displayed image that you want to scan, and the driver will set the scanner to read just that area. In essence, this is similar to making an enlargement of one section of a negative. You can also have the driver make color and contrast corrections to the original image. A sharpness menu allows you to set the degree of sharpness the program will add to the scanned image. You can choose "normal" high-contrast for line art, "exaggerated" for use with continuous-tone color photographs, or "soft" to eliminate halftone dots when scanning previously printed material from magazines.

In addition, film scanners are used to digitize slides or negatives, and differ from scanners designed to digitize prints in at least two important ways.

> First, these scanners are about the size of a slide projector, and work by passing light through a slide or negative into a fixed focus lens, which focuses the image onto a row of CCD light receptors which scan the image. With appropriate software (drivers), most scanners will allow different areas of the

image to be selected for digitizing. This is functionally equivalent to making an enlargement of an area of the slide/negative.

Second, most film scanners have a fixed resolution. Unlike prints that can be any size, all 35mm slides are the same size. Thus, a provision for selecting variable pixels per inch to allow you to fit the image to be scanned to the dimension of the screen is not provided. However, software may be included with the scanner, or may be bought separately, that will allow you to resize the resulting file, making it either larger or smaller. Following are a number of 35mm scanners suitable for digital photography.

The Barneyscan can make TARGA and Vista and compatible files with up to 16.7M simultaneous colors. It will convert 35mm color negatives into color-corrected positives, which will allow you to shoot negative film, skip making prints, and simply digitize the negative. The file created is 1520 pixels per row x 1024 pixels per column, for a total of 1,556,480 pixels. The resolution of 1520 x 1024 is below that of all the other film scanners listed below. However, included with the Barneyscan is software that will allow you to increase the size of the resulting file to a maximum of 3040 pixels per row x 2048 pixels per column, as well as reduce the image to lower resolutions. Software included with the Barneyscan also allows for automatic or manual exposure control; color, brightness, and contrast control (all of which can be viewed on your digital photography graphics monitor as you make the corrections); and cropping, sharpening, and conversion of a full-color image into 256 shades of gray. The Barneyscan lists for about $8,700, which includes both an expansion board and the driver needed to run it. A Macintosh II version lists for about $9,500, including the software and expansion board.

Another 35mm film scanner is the Howtek Scanmaster 35, which will create a TARGA, Vista, or compatible file with up to 16.7M simultaneous colors with a maximum resolution of 2000 pixels per row x 1333 pixels per column, for a total of 2,666,000 pixels. The Howtek Scanmaster 35 lists for $7,000. You must purchase Scan-It or some other driver to run it, as well as a GPIB, which might increase the cost by as much as $1,750 or more. Like the Barneyscan, it too will convert a 35mm negative to a positive image. A Macintosh II version is available.

The Kodak Eikonix 1435 will not only create a full-color file with 16.7M simultaneous colors, but one capable of a maximum of 4096 x 4096 pixels (which functionally is 4096 x 2736 in a 35mm ratio), for a total of 11,468,800 pixels in 35mm format. This number of pixels is suitable for reproducing the finest detail produced from the finest lens with the finest film available. It accepts 35mm slides and negatives, either mounted or in strips, and works with TARGA-16, 24, and 32 and Vista files. It lists for $9,900. Software, available from Eikonix (a Kodak company) for the TARGA boards costs about $495. A GPIB is also needed.

The Nikon LS-3500 Film Scanner also costs about $10,000. It can produce digitized images in 16.7M colors with resolutions up to 6144 x 4096 pixels, for a total of 25,300,992 pixels. It uses either a standard serial I/O port or the faster working GPIB. Drivers for TARGA boards are available, which allow for cropping, exposure and contrast control, color balance, creating high-contrast images from continuous-tone originals, and making posterizations and solarizations.

Digital Copy Cameras

Another way to scan a print is to use a digital copy scanning camera, which looks similar to a standard copy camera. The system consists of a baseboard with a vertical column extending up from it. Attached to the column is a device similar to a film scanner, consisting of a lens that focuses an image of the art on the baseboard onto a CCD chip. As the camera is moved further from the art, the scanned area becomes larger. The *Eikonix* 1412, from Eastman Kodak, is a digital scanning copy camera capable of creating a file of 4096 x 4096 pixels, (4096 x 2736 in 35mm film format with a total of 11,468,800 pixels) with up to 68,719,476,736 colors. It not only digitizes prints up to 40 x 40 inches in size, but it can also scan film as small as 16mm, and to some extent, 3D originals. This last point means that you can place objects on the base board, and as long as they are less than about 1 inch high, they can be digitized. The Eikonix 1412 costs about $27,000, and software to create TARGA-16, 24, and 32 and Vista files is available from Eastman Kodak (Eikonix division) for $495. A GPIB is needed.

Scanner Quality Versus Video Quality

You will have noticed that both video cameras and scanners use CCD chips. In the previous chapter I mentioned that the quality of the image produced by current video cameras suffers to some extent from a lack of tack sharpness and problems in maintaining highlight and shadow detail. In this chapter I've mentioned that scanners are capable of producing excellent images. If both use CCD chips, why is there such a disparity in image quality?

A complete answer to this question is beyond the scope of this chapter. Suffice it to say that (1) the type of CCD chips used in scanners differs in some important ways from those in video cameras, and (2) the CCD chip in the scanner sends its information directly to the computer where it is put into a file, while the information from the CCD chip in the video camera is transformed into a video signal and then sent to the frame grabbing function of the digital photography graphics board, where it is turned back into numbers. Thus, we have two very different systems, although both use CCD chips.

SOME PROBLEMS WITH SCANNING AND DIGITAL PHOTOGRAPHY

Depending on the scanner used, you can create a TARGA or TARGA-compatible file with grids as large as 6144 pixels per row x 4096 pixels per column (25,165,824 pixels), thus maintaining all the detail in the original image.

You will have noticed that I have mentioned not only the grid sizes the scanners produce, but also the total number of pixels the grid contains. For example, a file of 2000 x 1333 (the size of a file created by the Howtek Scanmaster 35) contains 2,666,000 pixels, and the file created by the Nikon LS-3500 (6144 x 4096) has 25,165,824 pixels. If these numbers represented the storage space in bytes required by the hard disk, it is easy to see that a 20MB hard disk could hold only seven files created by the Scanmaster 35, and would be inadequate to use with the Nikon LS-3500. It's not difficult to see that high-resolution files eat up a large amount of hard disk space. And neither of these cases takes into consideration software, such as your paint program(s), that must also be stored on the disk.

In reality, the situation is even worse than that sketched out above, because the number of pixels does not represent the storage requirements of the image file. In Chapter 2 I discussed *bytes*, and said that they can be thought of as any character or digit that can be typed into the computer from the keyboard. Each byte consists of 8 bits which can be thought of as light switches that can be either off (having a negative electric charge) or on (having a positive electric charge). It is the combination of 8 bits (some on and some off) that dictates the meaning/information carried by each byte.

In computer graphics, keyboard symbols aren't the essential units of information. Light and color are the relevant issues. The arrangements of bits in a file tell the computer graphics board what color to send to the monitor, rather than symbolizing the letter J or number 7. The TARGA-16 graphics board (and compatibles) use 16 bits (2 bytes) to create one colored pixel.

Therefore, a file created by the Howtek Scanmaster 35 to hold information for an image of 2,666,000 pixels would be 5,332,000 bytes in size in TARGA-16 format. If you created a digitized file from the Nikon LS-3500 generated image consisting of 25,165,824 pixels, you would have a file containing 50,331,648 bytes in TARGA-16 format. These files containing the image also contain some technical information needed by the computer program to enable correct operation, so the actual file sizes are slightly larger.

The problem of disk storage space is nothing compared to those problems encountered when actually manipulating large file images. If you used a Howtek Scanmaster 35 film scanner to create a high resolution 35mm ratio file of 2000 pixels per horizontal row x 1333 pixels per vertical column, the file would be about 16 times larger than 35mm ratio files the TARGA board is designed to work with: (2000 x 1333 = 2,660,000 and 510 x 340 = 173,400; 2,660,000/173,400 = 15.5). If you used the Eikonix 1412 to digitize a 35mm print at 4096 x 2736 resolution, you would have a file about 64 times bigger than what the TARGA board can handle.

This raises a dilemma. Large resolution files are desirable because of the quality of the image, but they are impractical because they are too big to be manipulated by the TARGA board. One solution to this problem is to work on these large files in sections. Using special software, discussed in some detail in the next chapter, it is possible to electronically break a large file into sections. Since a 2000 x 1333 file is about 16 times larger than a 510 x 340 file, (the size of a full-screen image in 35mm ratio), this large file can be broken into 16 sections, and each section can be individually manipulated by the TARGA board. Although this resolves the immediate problem and enables manipulation of large files, it introduces a certain amount of difficulty when the goal is to move objects from one location to another, or to lighten or darken the entire image. (The small rectangle in Figure 4.1 represents about 1/16 of the total picture area, which should give you a visual idea of what working on an image in 16 sections might entail.)

For reasons that will become clearer in Chapter 7, the most practical solution is to use a scanner to make files that can be addressed and displayed as one complete image on the digital photography graphics monitor. You can create (with most scanners) an image of about 510 x 340 pixels for a 35mm ratio file. The resulting scanned image will not contain all of the photographic detail in the original print or slide, but the image can be easily manipulated, which after all is one of the main reasons for digitizing the image in the first place!

SERVICE BUREAUS

A service bureau is to a digital photographer (or anyone working in computer graphics) what a professional photo lab is to a traditional photographer. Most service bureaus can scan slides or prints to any format you desire, for a fee which usually runs from $20 to $50. At this time service bureaus are not very common, and most work on relatively low resolution images of charts and graphs. Those set up to work on higher resolution TARGA and similar type images are as rare as hen's teeth. The chances are high that you will not be located near one, and so you may have to send the pictures to be scanned through the mail, along with disks to put the image on, and wait for the disk containing the scanned image to be returned. Most service bureaus will send your work back to you via some form of overnight delivery for an additional fee, and this can speed the turnaround time to a few days. While even this short turn-around time can be bothersome, especially if you are on a tight deadline, the savings in not having to outlay the cash for the scanner or other input equipment will make them attractive to many. As time goes on, more service bureaus will open shop, and eventually they will be as common as traditional photo labs. It will then be possible to drop off a slide or print, and return a few hours later for the scanned image file.

SUMMARY

In Chapters 5 and 6 you were introduced to various devices capable of putting a photographic image into your computer: video cameras, still video cameras, audiovisual, and scanners. Consumer grade video cameras offer the advantage of being inexpensive, but the quality, with regard to both sharpness and detail, may be inadequate for your needs. Industrial grade cameras, especially three-chip RGB cameras, yield higher quality images, but the short tonal range, as well as the high price of these cameras, will rule out this option for many. B&W video cameras do usually offer higher quality for a lower price, but limit you to working in black and white. B&W scanners offer high quality at a relatively low price, but also limit you to working in black and white. Color scanners offer excellent quality, even at 512 x 342 resolution, but at a high price. When scanners are used to create high-resolution files of 2K or longer, the files eat up disk space, and force you to manipulate the image in many sections.

Although there are problems associated with all of these input devices, scanners produce the highest quality images, and so I recommend that you investigate scanners for your input device. However, using a home video camera when first getting started will lower the initial cost of your digital photography system, while giving you the opportunity to become familiar with all the aspects involved in digital photography.

As my suggestion for the type of scanner you should buy, I suggest none, at least at first. Try having some prints or slides digitized by a service bureau. I suggest you try two different service bureaus, and compare their work: Pick two similar slides or prints, and send one of them to one service bureau and the other to another; compare the quality of these scanned images to each other, and to the quality of the video-grabbed images you are used to working with. If one of the scanned images seems superior to the other (e.g., truer colors, correct "exposure," the image is correctly oriented and not tilted to one side), then stay with that service bureau for your scanning needs.

After you have some sense of the advantages and disadvantages of scanned images for your purposes, decide whether your business can function with the delay in sending off the slide to the service bureau, and waiting for the scanned image file be returned to you. If the type of work you do has deadlines of two weeks, you may decide to put off buying a scanner until your digital photography business has enough clients to justify the expenditure. If you have deadlines of a few days, then you probably will need to get a scanner of your own.

If you feel that you need your own scanner, decide whether a gray-scale scanner will meet your needs. If it will, I recommend you look into a Hewlett-Packard ScanJet Plus. With its variable resolution of 12-1500 dots per inch, and its 256 level gray scale, it is a very good buy. You may also wish to investigate other gray-scale scanners your digital photography reseller recommends. Keep in mind that you may have to buy a driver to work the scanner, and probably a conversion utility program to change the non-TARGA-type file to a TARGA-type file.

If you decide you need a color scanner, the safest choice is the Howtek Scanmaster II for prints, or the Scanmaster 35 for 35mm slides. Safe, because Howtek/Sharp scanners are probably the scanners most commonly used by small businesses and individuals. While you will not be breaking new ground, the ground you are traveling has been well mapped out for you by previous buyers. The $6,750 to $8,750+ price (including software and GPIB) does buy a proven product which will yield extremely high quality results. The Barneyscan, for about $8,700, which includes an expansion board and software, is also an excellent scanner, and costs about the same as a Scanmaster 35, once the GPIB and

CHAPTER 7

PROGRAMS (MAGIC)

So far we have discussed the base computer, various types of graphics boards suitable for digital photography, and suitable methods to get an image into the computer. In this chapter we will get down to the *fun* part of digital photography: manipulating the image.

This chapter is divided into seven parts. First I will introduce two pieces of equipment, the mouse and digitizing tablet, which are used to move the *cursor* around the computer graphics monitor in order to identify and isolate the parts of the image you want to manipulate. Then I will discuss, in general terms, *paint programs*, which are used to manipulate the image. I will describe in greater detail three programs, TIPS, LUMENA, and RIO, which are the *standard* programs I mentioned in Chapter 4. Included in some of these sections will be brief discussions of other functionally equivalent programs. Some examples of how you would use these programs to create a number of effects will be given. *Utility programs* will be presented which are used either to enhance these programs or to enable you to do some things the standard paint programs do not do. Finally, a summary in which I will make suggestions about which program(s) may be of most value to you, concludes this chapter.

MICE AND TABLETS

Technically speaking, mice and tablets are both *input devices*, in the same way a computer keyboard is an input device. All take information from outside the computer (you) and put it into the computer. After the keyboard, the mouse is probably the most popular method for giving instructions to a computer. Its name comes from its remote resemblance to a mouse (Figure 7.1). There are numerous mice on the market, and most have list prices between $75 and $150. Because these devices can be used with many different types of programs, there is a large demand for them, and so there is much competition among the various manufacturers to get you to purchase theirs. Often a mouse is *bundled* with some software, such as a paint program for EGA/VGA use, as an enticement to buy a particular brand. Mice are often discounted by mail order houses.

The mouse that set the standard for how mice should operate is the Microsoft Mouse, and many other mouse manufacturers include an *emulation* mode which will make their mouse act as if it were a true Microsoft Mouse. Because the prices of mice are within a small range, and because almost all paint programs recommend a Microsoft Mouse, I suggest (and will assume) that you get the Microsoft Mouse.

The Microsoft Mouse comes in two different models which are equivalent in appearance and function but differ in how they are attached to the computer. One attaches to a serial I/O port and the other, called a *bus type*, comes with an expansion board which fits into an expansion slot on your motherboard. Neither type is superior to the other. With one you take up an expansion slot, and with the other you lose the use of a serial port. The choice is yours. If you buy your computer and mouse from a Truevision or Vision Technologies reseller, the mouse will be installed for you. If you buy it from a mail-order company, you will probably have to install it yourself. Complete and easy to understand instructions will come with it, and you can get help if needed from the customer support line of the mail-order company, from Microsoft, from a friend, or from a computer repair shop.

As is shown in Figure 7.1, the Microsoft Mouse has two buttons at the front. These buttons are flush with the top side of the mouse; that is, they don't stick up. If you are right-handed, your forefinger will rest on the left, slightly larger

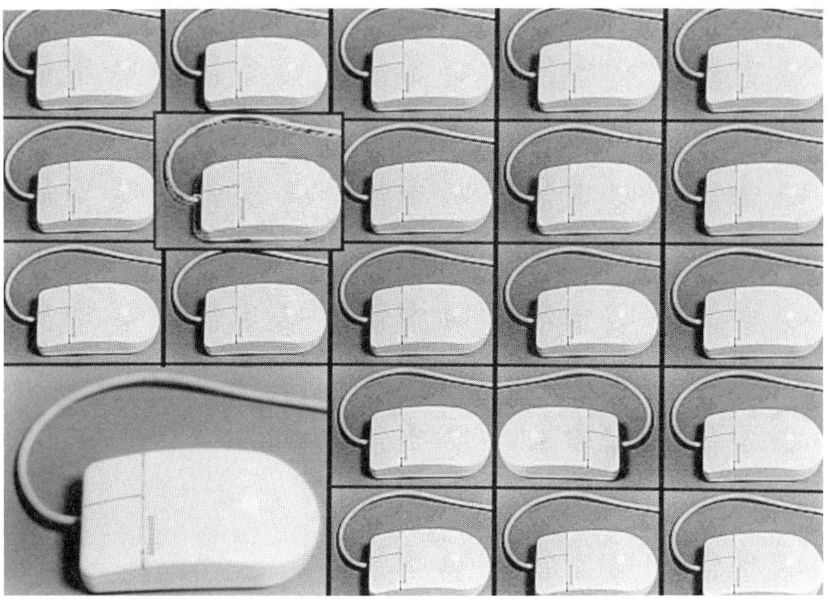

Figure 7.1 A mouse. Image made with a Canon E-708 8mm video camera, TIPS and QFX utility program.

button, and your middle finger will rest on the right button. Your thumb will grasp the right side of the mouse's case, and your other fingers will curve around the other side. The palm of your hand will rest toward the back of the mouse. On the bottom of the mouse is a round, hard, rubber ball, mostly inside the mouse, but protruding slightly from it. You move the mouse by sliding it on the surface of your desk. You may want to purchase a mouse *mat* to slide your mouse on. They are made out of hard rubber, are slightly textured, and cost between $5 and $10. Because they have a rough texture, they increase traction, and thus make the ball rotate easier. This is purely optional, since the mouse will work well without one.

As you slide the mouse around, the hard, rubber ball rotates. There are sensors inside the mouse that react to the movement of the ball. These sensors translate the movement of the ball into signals which in turn move the cursor on your graphics screen. The cursor, as mentioned earlier, shows you where on a monitor the next instructions from the computer will take place.

If you have never used a mouse before, you're in for a giggle. The instructions are straightforward. To move the cursor toward the left of the screen, you slide the mouse to the left of your desk. To move the cursor toward the right, you slide the mouse to the right. If the mouse were positioned in front of you, and you wanted to move the cursor toward the top of the screen, you would move the mouse away from your body, and to move the cursor toward the bottom of the screen, you would move the mouse toward your body. You can move the cursor diagonally, or in a circle, or randomly around the screen, just by moving the mouse in the appropriate direction(s) on your desk. You will find no problems in moving the cursor left or right, but moving the cursor up, down, and diagonally takes some getting used to.

The eye-hand coordination involved in trying to move something *up* by moving something *away*, or moving something *down* by moving something *toward* you is not intuitive, and takes some practice. And getting the cursor to move diagonally (both up or down and to one side simultaneously) is almost impossible at first. The first few hours of trying to get the cursor to go where you want it might involve a lot of body English, which will do absolutely no good at all. Nevertheless, you will probably find yourself leaning in various positions trying to guide the mouse. Don't get frustrated. Instead, enjoy the very weird feeling of trying to do something that is simple in theory, but isn't in practice. After a few hours you will begin to get the hang of it, and after a few days, you will have trained your brain to accept the fact that you must move your hand to the left and away from you to move the cursor to the left and up. At this point, using the mouse will become routine, and you will give it little, if any, thought.

For the most part, you will not be depressing either the left or right buttons when moving the mouse around. You use the buttons when the mouse is in the correct position and you want to give instructions to the computer program you are using. The mouse is an excellent and inexpensive tool for use in digital photography, and we will talk more about the buttons and their use later in this chapter.

Besides getting used to manipulating them, the only problem with mice is that they are not particularly good for drawing. If you want to make drawings and combine them with digital photography, or if you are primarily interested in animation or some other area in which a lot of drawing is necessary, you will be better off with a tablet. A tablet looks similar to a clipboard. Attached to the tablet by a thin wire is a pen-like *stylus*, and a cable connects the tablet to a serial port in your computer. Inside the tablet are a series of thin wires arranged in a grid-like pattern. When you touch the tip of the stylus to the tablet, a signal is transmitted to the wire grid, which positions the cursor on the graphic's monitor. As you move the stylus on the tablet, the cursor will move on the screen. Because the stylus has the look and feel of a pen or pencil (although it is a bit thicker than the typical pen or pencil), it is easier to do freehand drawings with it than with a mouse. If you pick a mouse up off the desk top and place it in another position, the position of the cursor won't change, since the ball won't have rotated. But you can move the cursor to a different position by lifting the stylus from the surface of the tablet and then putting it down in a different position.

Most tablets come with or can use a number of optional styli. Some have ink cartridges which enable you to place a piece of paper on the tablet and draw on it while simultaneously moving the cursor on the screen. While it may seem that this would be a necessity, it isn't since you will rapidly get used to watching the computer graphics monitor while you move the stylus around on the tablet. For this reason, most people use a stylus that is similar to using a pencil or pen without a point. Other styli are similar to using a mouse. All styli have two or more slightly raised buttons near their tip, which are equivalent to the buttons on the mouse.

Some tablets are pressure sensitive, that is, they will change the look of the drawing on the graphics monitor depending on the amount of pressure you apply when drawing. For example, if you were drawing a straight line, you could make it thicker at some point by pressing down harder on the tablet, and then make the line thinner by easing up on the pressure. If you plan to do a lot of drawing, a tablet of this type may be of value, but the paint program you use must be able to interpret these signals. Buying such a tablet won't do you much good if your program won't support this feature.

Tablets come in different physical sizes, which means that the drawing area of some is larger than that of others. Since a larger drawing area necessitates a larger and more intricate wire grid, larger drawing tablets cost considerably more than smaller ones. For digital photography purposes, most tablets cost somewhere between $450 and $600. Tablets, as with mice, can be bought from the same people you get your graphics board from or from a mail order house. Just as the Microsoft Mouse set the standard for mice, Summagraphics set the standard for tablets.

PAINT PROGRAMS

A paint program takes instructions from you and translates them into a form the computer can understand. The purpose of this type of program is to create and manipulate images. They are called paint programs because they were, and still are, designed for use by graphic artists to draw or paint images, and not primarily for use in digital photography. Happily, the effects the paint programs allow you to create are useful in digital photography. All paint programs include the following three types of functions that are used in digital photography.

The first type of function is housekeeping chores, which includes creating, maintaining, and deleting files of images, as well as other standard computer functions that are necessary for you to integrate what you do with the program with the standard functions of your computer. These are not creative in nature, just run-of-the-mill computing functions.

Secondly, paint programs provide effects that emulate basic darkroom and shooting procedures. These may include the equivalent of opaquing a negative; spotting or retouching a negative or print, including the use of airbrush and hand-coloring techniques; and the ability to change perspective, which is similar to using shifts, swings, and tilts on a view camera.

Finally, effects are possible that are very difficult, if not impossible, to achieve with traditional photography. Some of these effects include the ability to reverse (flop) a *portion* of the image, while leaving the remainder unchanged; moving a portion of the image to a different location; enlarging or reducing in size a portion of the image; removing selected areas and replacing them with information from another photograph or drawing; and distorting portions of images, to name just a few. With these and other effects, the photographer has far more control than ever before, which leads to new ways of thinking about photographic images. Once you have learned to use these features, you will never approach photography in the same way that you have in the past. Moreover, these features can make standard photographic effects very easy to accomplish.

For example, think of photographing a bottle of wine. The client wants the cork to be floating a few inches over the top of the bottle. There are at least two ways to achieve this effect with traditional photography. First, you can try to suspend the cork over the bottle. You could put a needle into the cork, and attach monofilament (fishing line) into the needle. You will probably find that the dangling cork tilts slightly to one side, or is difficult to get centered directly over the bottle. With some time and patience, you should be able to get the cork oriented correctly. Depending on your lighting, you may pick up a slight reflection on the monofilament, and thus have to have this area airbrushed. A second method would be to shoot the bottle without the cork, and then make a second shot of the cork, and have the lithographer superimpose the cork into the picture. This will add to the cost of the job, and the cork may look as if it were pasted onto the original photo.

With digital photography, there are a number of ways to float the cork above the bottle. The simplest would be to place a cork so it is just inside the neck of the bottle. If you shoot the scene with standard film, you could scan the resulting slide, negative, or print, creating a digitized image. Using the paint program, you could then isolate the cork and move it to a position above the bottle. (More about isolating sections of a photograph will be discussed later in this chapter.) This should take about 15 seconds. You may notice that the bottom of the cork looks slightly curved, since part of it was inserted into the bottle. You can instruct the program to "remember" the color and texture of the cork. Then, using the free-hand drawing commands, you could straighten out the bottom of the cork, making additions that match the original exactly in color and texture. If you have trouble drawing a straight line, you can tell the computer to do it for you. Using the mouse or tablet, you tell the computer program where to start the line, where to end it, and the computer will take care of the rest. This entire operation may take less than a minute (once you have become accustomed to using the needed commands).

This basic technique can be used to float any object. The computer "remembers" the numbers associated with the location and colors of the object to be moved, and then rewrites them to a different location in the image. You have the option of leaving the original object in place, thus creating one or more duplicate objects, or *cutting* the original object out. If you do this, a white area will appear in your image. If the wine bottle was shot on a plain seamless background, this is no problem. The cork will be moved, and a white area will appear where the cork was originally located. You can tell the computer to *read* the color of the seamless background surrounding the now white area, and to *fill* in the white area with that color.

At present, three major paint programs are used in digital photography. These can be thought of as *standard* programs because they are so widely used. Some of the major features of each will be presented.

TIPS

TIPS (for *T*ruevision *I*mage *P*rocessing *S*oftware) is a paint program written by Island Graphics. It is probably the first paint program many of you will use for digital photography. It was designed for use with the TARGA series of boards and different versions are needed for each board. TIPS for a TARGA-16 won't work with a TARGA-32, nor will it work with the Vista 4M. Another version of this program, called *FreeStyle*, which is functionally the same as TIPS, has been designed for the Vision-16 and 32 boards.

The reason you will probably begin with TIPS is that it is now *bundled* with TARGA-16 and M8 boards, although in an incomplete form. The incomplete (free) version will allow you to create many of the effects that are available in the complete version, but not all. The marketing strategy is that if you get it free, you will use it, become familiar with it, and then purchase the full version rather than a competing program. This marketing strategy was begun in 1989, and if it is deemed a success, TIPS will continue to be bundled with the TARGA-16 and M8 boards.

TIPS comes with an easy to understand, well-organized user manual. If you buy your computer and graphics board from a Truevision (TARGA) reseller, they will install it for you (i.e., put it on your hard disk and make sure everything runs correctly). If you have the TARGA or Vision board sent to you, along with TIPS or Freestyle, you will have to install the program yourself. Complete instructions are in the manual, and your reseller is only a phone call away if you run into difficulties.

To run TIPS with a TARGA-16, you will type TGATIP16 <cr> on your computer keyboard. (You can use a MSDOS program called *rename* to change TGATIP16.COM to any other legal name. For example, you can rename it T.COM. If you do, you will just have to type t<cr> to start TIPS instead of the longer name.) A few seconds after you instruct your computer to run the program, the program will display the name TIPS on your color graphics monitor. Push the left or right button on the mouse. The screen will clear and the main menu will appear. The main menu will cover about 20% of the graphics monitor screen. It has 26 small icons (cartoon-type drawings). The first row of five icons represents drawing functions, the second row represents special functions, the third row allows you to select the size and shape of cursor to be used, the fourth row depicts special patterns you may wish to use for backgrounds or other purposes, and the fifth contains more special functions.

This main menu will be replaced by other menus. For example, if you want to see a complete selection of all the drawing functions that are available, you would move your cursor (by moving the mouse) so it is on the first icon in the first row. This icon shows a pencil and a t-square, and is used to symbolize drawing functions. When the cursor is on the pencil and t-square icon, you press and release the larger left button (also called the select button) on your mouse or the equivalent button on a stylus. The menu will now change and show icons representing all the drawing functions available in TIPS. On this new menu are symbols for creating a rectangle, circle, ellipse, trapezoid, curve, polygons, and arcs, to name just a few.

To create a rectangle, move the cursor to the icon depicting a rectangle, and press and release the select (left) button. A new menu appears, with six choices. Some of these choices can be used in combination with some of the other choices, but some of the options are mutually exclusive. One of the choices is "filled." If you place your cursor on this word, and press and release the select (left) button on the mouse, the word will become highlighted (brighter). This indicates that you have selected this feature. If you now press the cancel (right) button on the mouse, the main menu will again appear, but this time with the rectangle icon highlighted, indicating that the program is currently set to create rectangles. Press cancel again and the menu will disappear and the screen will be clear. You can now move the cursor anywhere on the screen. When you press and release the select button, you will have selected the location of one corner of the rectangle you want to create. As you move the cursor around the screen, you will notice that a rectangle appears and continues to change shape as the cursor is moved. One corner of the rectangle is fixed at the place you indicated when you began to create the rectangle, the diagonally opposite corner moves as you move the cursor. You can make any size rectangle you wish, up to the maximum size of your screen. It can be long and narrow, or perfectly square. When you have made the desired outline, press and release the select button, and a rectangle *filled* with the color the computer is currently using, will appear. If you want to create another rectangle, simply repeat the process.

If you want to change the color, or use another function, press cancel. The main menu will reappear superimposed over the rectangle. On the bottom of the main menu is a small square of color, and below it is a color chart, called a palette. The *default* color is red when TIPS first comes on, and so the rectangle we just made will be filled with that same red. To select another color, you would move the cursor to the color on the palette you want to use, and press and release the select button. The color square on the bottom of the main menu will change to the new color. This square will always show the color the computer is currently using. To create another rectangle with the new color you just chose, press cancel. The menu will disappear and the graphics screen will show the previously drawn rectangle. Move the cursor to where you want one corner of the new rectangle to appear, press and release the select button, and then move the cursor to create the size and shape of the new rectangle. When you press and release select again, a new rectangle will appear, this time in the new color. It is possible to place all or part of the new rectangle over the rectangle previously drawn. The area of the old rectangle which falls behind the new rectangle will appear in the new color.

If you want to create circles, press and release cancel to go back to the main menu, move the cursor to the pencil and t-square icon, and press and release the select button. The menu of drawing functions will appear again. Move the cursor to the circle icon, and then press and release the select button. A number of choices will appear. This time don't select any of the options. Instead, press cancel to get back to the main menu, which will now highlight the circle icon, not the rectangle. Then press cancel again to remove the menu from the screen. Now you can move the cursor to a position on the screen where you wish the center of the circle to be, and press and release the select button. As you move the cursor, you will notice a circle appear and get bigger as you move the cursor away from the center of the circle. When you have the size of circle you want, press and release select and the *outline* of the circle will appear in the color the computer has been set to use. This circle is not filled in, because we did not choose the *fill* function from the circle menu. If we wanted to have a solid circle, then the fill option would have had to have been chosen, as it was with the rectangle example.

Most paint programs work in a similar manner. You choose the effect you want from a menu, set any parameters, such as fill, and then use the mouse to position the function or effect. The TIPS manual has a large "step-by-step" section that describes in detail how to create circles and other geometric shapes, as well as how to use the different functions and effects.

You may be wondering how creating a rectangle or circle relates to digital photography. There are two reasons to learn how to use these features. First, if you want to isolate a cork in a bottle of wine and move it to a new location, you will have to draw a rectangle around the cork to isolate it from the bottle. Thus, learning how to make a rectangle has practical photographic applications. Second, you can use these geometric shapes to rapidly "opaque" (blacken out) areas of the image, to use with other functions. If you wanted to opaque a large portion of sky (to later replace it with a sky from a different photograph, for example), you could use the freehand drawing tools to paint black over the sky, or you could use a filled rectangle to rapidly cover a large portion of the sky.

When you first start using TIPS (or another paint program), I suggest you spend a few days making designs with the geometric shapes and other features of the paint program. Don't get frustrated if you are not creating great art. You are learning how to use the tools you will need for digital photography. Some of the functions in TIPS that are important in digital photography are outlined below.

The *move* function includes a number of different functions. With it, you can draw a rectangle around an area of the screen, and then reduce or enlarge it; you can stretch or shrink the horizontal, vertical, or both horizontal and vertical dimensions; you can rotate the selected image up to 360°, flop it left to right or top to bottom, or move the area to a different location in the image.

The *slide* function allows you to move part or all of the image vertically, horizontally, or diagonally. This is similar to using the front and rear rise, fall and shifts of a view camera. Unlike a view camera, which moves the entire scene, the slide function will allow you to move any part of it. This is handy if you want to line up two or more areas of the image. For example, if you want a cloud to be directly over a house, you can isolate the cloud and slide it left or right so that it is in the desired position. If you want the cloud to be higher or lower in the sky, you can use slide to reposition it. This same effect can be achieved with the move function, but sometimes it's better to slide part of the image than to move it. Experience will tell you when one function is better to use than the other.

The *dupe* (for duplicate) function allows you to add or copy a part of the existing image to another region of the image. For example, if you have a lopsided tree in your image that you want to make more symmetrical, you can tell the computer to "remember" the branches in one location and then, using the drawing function, position the cursor to where you want the new branches and they will appear. By adding branches that have been duplicated from different parts of the original image, you can create a tree which, although built from the original tree in the image, has branches that are not identical to the branches in the original tree.

The *grab* function is used to capture an image from a videotape or a live video image. You can digitize the image by pressing the select button on the mouse (functionally the same as pushing the shutter release button on a camera). With this function you can also adjust the hue, contrast, and intensity of the image. You can use these functions to color correct the image (the functional equivalent of using color filters on an enlarger for color correction), create overall tints, or increase or decrease the contrast of the image. The hue, contrast, and intensity control only work with video images, and cannot be used to change the appearance of images once they have been scanned or otherwise digitized.

The *tint* function allows you to replace one color with another, while maintaining the photographic texture in the original. For example, you could use *tint* to change the color of someone's eyes from blue to green: Select the color green you want and then draw over the blue in the model's eyes, using the drawing tools; the blue will be replaced by the green, but the photographic texture of the eyes will be retained.

Perspective is a function that allows you to control the perspective in your image. It is similar to using the front or back tilts or swings on a view camera.

Thus, you can correct for distortion (parallax) in your original image, or create distortions if you wish.

The *tile* function allows you to make an overall pattern out of part of your image. Let's create another hypothetical image. Let's say you photograph a rose. You could scan it and then use the move function described above to reduce it in size to about 1 square inch on your graphics monitor screen. You could use the tile function to "remember" the small rose. Now you could choose *filled rectangle* and use the cursor to create a rectangle the size of the monitor screen. Instead of filling the screen with a color, the screen will be filled with numerous 1 inch roses. The effect would be that of rose-patterned wallpaper. You could then place one or more objects in front of this rose background.

.Win files (short for *window*) are special files used for merging two or more images together. Lets assume you created the rose-patterned wallpaper as just outlined. You could *save* it in a paint file called "rose" by telling TIPS you want to store the image that is on the graphics monitor in a file. You will be asked to give the file a name. You type in *rose* and TIPS takes care of the rest, including adding the three letter file extension *.tga*. This extension identifies a TARGA paint file used by TIPS. Let's say you had previously digitized an image of a model, and had stored it in a TIPS paint file called *model*. To call this image to the monitor screen, you tell TIPS you wanted to *load* a paint file. TIPS displays a list of all files ending in .tga. You place the cursor on the file named *model* and press and release the select button. The image of the model appears on the screen, replacing the image of the rose wallpaper. Using the drawing tools, you now isolate the image of the model by choosing the black color on the palette, and painting over the background and any other area of the image that is not the model. When you are finished, you have an image of the model surrounded by black. This procedure is functionally the same as opaquing a negative. You then save this new image as a *.win* file by telling TIPS you want to save the current monitor screen as a window file. TIPS will direct you to draw a rectangle around the area you want to make into the window file. This differs from saving a paint file. All paint files contain the entire information on the computer graphics monitor screen when the file is created. A window file can be of an entire monitor screen, or any rectangular portion of it. If we are only interested in the model, we only need to draw a rectangle around the model. After you create the rectangle, TIPS asks you to give a name for the file. Let's say you type in *model*. TIPS creates a window file called model.win and stores it on your hard disk.

Notice that we now have two files called "model". As I mentioned earlier in this book, two files with the same name are not permitted on the same disk, because the computer cannot distinguish between them. However, our two "model" files are different, because they have different file extensions. One, the original paint file, is labeled "model.tga," and the window file is labeled "model.win." Because these files have different file extensions, the computer can distinguish between them.

Now lets load the "rose" file so it is on your graphics monitor screen. To do this, you tell TIPS you want to load a paint file. All the .tga file names appear on your monitor. You place your cursor on the one named *rose* and press and release the select button. The image of the rose wallpaper appears on the monitor screen, and the one of the model disappears. Finally, you load the *model.win* file. This time, you tell TIPS you want to load a window file. TIPS displays the names of all the files that end with the extension *.win.* You place the cursor on the name *model,* press and release the select button, and the image of the model appears on the screen. But this time, the image of the wallpaper does not disappear. Instead, the screen will now show both the model and the wallpaper. The image of the model will cover up any part of the wallpaper the model is in front of, and the wallpaper will show through any area of the model image that is pure black, or that is outside the rectangle that makes the .win file. (Don't forget, a .win file can be smaller than a full monitor screen.) The effect will be of the model standing in front of the wallpaper. At this point you will have the option of moving the model around the scene. If the image of the model is in the center of the picture, and you want to position it to the side, you will place the cursor on the model, and while pressing down on the select button, move the mouse to reposition the model. When you are satisfied with the composition, you save the new image in a paint file, with a name such as *finished.*

One may wonder what would happen if the model had black hair or was wearing black clothing. You will find that pure black is rare in a digitized image. Black clothing usually appears in a .tga file as a number of very dark colors, including dark browns, reds, greens, blues, magentas, etc. Intermixed with these very dark colors may be a few pure blacks. These few black pixels will be replaced by the "wallpaper" image. Actually, these pixels will appear as isolated random color specks from the rose wallpaper. In this case, you would use the drawing tools to change any of the offending pixels to a pure black or to a very dark color, usually the same color as, or a similar color to, that of the adjoining pixels.

The *mask* function allows you to cover up an area of the image so as to make it impervious to the effects of some other functions. For example, you could use the mask function to black out the background of the *model file* described above. To do this, select the mask function from the menu, and then use the drawing tools to draw over the model. The model is not covered with a color, but with a translucent textured pattern, identifying which areas have been masked. When the model has been completely covered, choose *filled rectangle* from the menu, select pure black as the color, and then create a rectangle the size of the computer screen. The computer will make everything black except the area protected by the mask.

The functions listed above are only a few that TIPS has to offer. Since many of these functions interact with other functions, there are probably thousands of combinations of functions possible to select from. The TIPS manual is quite good, so you should have no difficulty learning how to use the functions, and after a few weeks you probably won't have to refer to the manual very often.

TIPS is not available for the TARGA-8, nor the Vista 1M and 2M boards. TIPS for the TARGA-M8 lists for $995, for the TARGA-16 $795, the TARGA-24 $1,250, the TARGA-32 $1,250, and for the Vista 4M $2,500.

While a modified version of TIPS is presently included with the purchase of a TARGA-M8 and 16 board, a program called ColorScheme I is included with the Vision Technology Vision-16 board. ColorScheme I is a basic program, and does not include many of the functions you will need, although it will allow you to get started. Also available for the Vision family of graphics boards is FreeStyle, which is designed by Island Graphics, the same people who designed TIPS. For all practical purposes, Freestyle and TIPS are the same program, although FreeStyle must be used with the Vision boards. FreeStyle-16 for the Vision-16 lists for $695, and FreeStyle-8 for the Vision-8 lists for $495. TIPS also works with the Artist RealVision-16 and 24 graphics boards. TIPS is not copy protected.

LUMENA

If you never use any other program but TIPS (or FreeStyle) you will probably be quite happy, since it will give you far more control over your photographic image than traditional photography does. Only after using LUMENA for a while will you know what you have been missing. LUMENA allows you to do all that TIPS does, and much more. Many of its features are designed for those interested in drawing or painting, and these functions (called *tools* in LUMENA) will not be of paramount interest to most photographers. However, LUMENA does have a host of tools that are of great interest to photographers. Following are a number I feel are the most important.

The *sharpen* tool allows you to sharpen all or part of an image. This is indispensable if you use an industrial grade video camera, a Canon or Sony still video camera, or the Fotovix II-X to create your primary digital image (but is less helpful if you start from a video image from a consumer grade video camera, since most of these images will be of too soft a focus to benefit much from sharpening). This tool also allows you to sharpen scanned images, if needed. The *sharpen* function allows you to choose from a sliding scale (1 to 100) the degree of sharpness desired. Over-sharpening will result in sharpening *artifacts*. Such an image would have an unfortunate resemblance to enlarged grain or dye found in color negatives or slides. If you do oversharpen the image by selecting too high a degree of sharpness, you can reload the original image and sharpen it again, this time selecting a lower value.

This last point warrants some discussion. When you load a file to work on, you are only making a copy of that file from your disk into RAM. As you manipulate the image, you are manipulating RAM. The original image file is untouched. If you do something you don't like, you can go back to the original image file and start over. As you manipulate the RAM image and make changes you do like, you should periodically make the image-in-progress permanent, to save you from having to start at the very beginning should you do something that creates an effect you don't like, or in case you have a power failure which would essentially destroy all work in progress in RAM. As you are working, periodically instruct your paint program to create a file of the image that is currently on the screen. You can give the file the name *working1*, and later store another version called *working2*, etc. If at some point you look at your image and decide you don't like some of the changes you have made, you can go back to any of the intermediate working files and start over again at that point. Once you have a satisfactory finished image, and you no longer need these intermediary files, you can erase them from your disk.

LUMENA also has tools that allow you to lighten or darken the image, as well as increase or decrease the overall contrast of the image. You can color correct an image by increasing or decreasing the relative amounts of red, green, and blue. These features differ from the Hue, Saturation, and Intensity functions in TIPS, which only work with the *grab* function (capturing an image from video or videotape). With TIPS, once an image is grabbed, the contrast, brightness, and color are pretty well fixed. With LUMENA, an image can be color and contrast-corrected whenever you choose.

.Win files in TIPS allow you to merge one image with another by covering up part of one image with the other, as was the case with the rose wallpaper and model example. LUMENA not only allows you to merge images, but to *blend* two or more images together, yielding a result similar to that obtained by using the "sandwiched negative" technique in traditional photography. In the sandwiched negative technique, a negative of a sky could be placed in contact with a negative of a field. When printed, the sky and field will be merged together. With LUMENA, the photographer is given complete control over the process. You can choose the degree of merging, from almost none to extreme. You can also decide where the merging should take place and where it shouldn't. Color Plate 5 shows an example of how this effects looks. The bricks in the wall were allowed to bleed through the photograph of the house, making the house look as if it were painted on a wall. The bricks could have been allowed to come through the image to a greater or lesser degree, thus altering the effect.

LUMENA also has a number of tools that give somewhat unpredictable results, depending on the makeup of your photographic image. Sometimes these effects work well, and other times, not so well. One, called *colorize*, can change a full-color image into one that looks as if it had been drawn with colored chalk. Another, called *emboss*, reduces a full-color image to two colors, giving parts of it a slightly raised appearance, as if it were printed and embossed onto a piece of paper. And another, called *threshold*, will create a high contrast effect, allowing you to decide at what level of intensity the image should go white (or another color) and at what level it should go black (or another color).

LUMENA has a built in posterization effect, with four levels of intensity. You can choose light, moderate, strong, or very strong posterization for all or part of a scene.

All of these tools appear on a series of menus. Unlike TIPS, LUMENA's menus use not icons, but words to denote your choices, and the menus appear on your standard monitor screen, not the digital photography graphics monitor. Also unlike TIPS, the menu in LUMENA can be modified to suit your purposes. If you use some tools often, you can group them together and have them displayed when you turn LUMENA on. You can also save a current working setup.

This means that if you have set certain parameters for different tools, and want to end the session, you can save the settings so that they will appear, already set, the next time you turn LUMENA on.

One fun feature is *animate*, which creates a file of all your actions when making a particular image. It is possible to play back this file, which will remake the image very rapidly, starting from the beginning. While it is not a particularly important feature, it may impress a client, or remind you of how you created a certain effect.

LUMENA is a modular program. This means that it is sold in sections. All of the tools just mentioned are found in the base module. Possibly in an attempt to counter TIPS marketing strategy of supplying a working but incomplete version of TIPS with each purchase of a TARGA-M8 or 16 board, LUMENA has added two modules, which used to be sold separately, to its base module. One is *Scan-In*, which will allow you to control a variety of scanning devices. Also included in the base module is *HI RES*.

In Chapter 4 I mentioned that there were ways to increase the maximum resolution of TARGA and TARGA-compatible boards to be equal to or greater than those of the more expensive Vista boards. And as we discussed in the previous chapter, scanners can create TARGA files with resolutions larger than 512 x 482 pixels. One problem with these large files is that you cannot display them in full on your computer monitor, since the largest TARGA image you can work on is 512 x 482 pixels. Scanning an image larger than this is counter productive.

LUMENA'S HI RES allows you to use scanned images up to 2K (2048 pixels per row) resolution. When using HI RES, the entire image can be viewed at one time. This is done by subsampling (also called *proxy editing*). Subsampling means that some of the pixels from a large image (over 512 x 482) are left out of the image on the screen. For example, a 2K image is four times wider than an image made up of 512 horizontal pixels. In order to show the 2048 pixels in each row on a monitor controlled by a TARGA board which is capable of displaying only 512 horizontal pixels, three out of every four pixels are not displayed. The first pixel in each row is displayed, but the second, third, and fourth aren't, then the fifth pixel is displayed, etc. If you imagine that leaving out so many pixels presents a grossly inferior image, you're wrong. The subsampled image is exactly the same as a scanned image of 512 resolution, and will look quite good. True, you have less detail than in a large scanned image, but your eye tends to put some of the missing pieces into the picture. While this subsampled view does not contain all the detail that is in the entire file, it does show you the overall composition.

HI RES places a grid over the subsampled image on your digital photography graphics monitor. At 2K resolution, the picture is divided into 16 rectangles. A small rectangle, 1/16 the area of the screen and that represents one full screen, can be moved anywhere within the picture area. (The small rectangle in Figure 4.1 represents 1/16 of the image area of a 2K resolution file, while the larger rectangle represents 1/4 of the image area of a 1K file.) You can therefore isolate any 1/16 area of the image, select it, and the program fills the screen with that selected area. This new screen contains all the information in the file for the part of the image you choose. It is no longer subsampled, but contains *all* the pixels for this portion of the image. You can make any changes you wish, and then instruct HI RES to paste the section back into the original file, either exactly where it came from, or if you choose, in some other location. You can then view the entire subsampled image to see how your changes affect the entire image, choose another section to work on, or call back the section you just finished to make some more modifications.

In addition to allowing you to use a large-scanned image up to 2K resolution, HI RES will create a larger image from an originally smaller file. In Chapter 4 I also mentioned that there were software means to increase the resolution of a 512 resolution TARGA or TARGA-compatible file to that of a Vista file. LUMENA's HI RES is one of these methods. For example, let's say you shot your original image with a Canon RC-760 or Sony ProMavica camera. The image, when made into a TARGA computer file, is 512 x 482 pixels. You could instruct HI RES to expand the image to either 1K, 1.5K, or 2K resolution. Let's assume that you wanted a 2K resolution file. Since you have 512 horizontal pixels and you want 2,048, HI RES must create three pixels for each pixel in the original image. We will call these new pixels *pseudopixels*, since they are created by the computer from existing pixels, but are not part of the original image. HI RES will give you two choices for the expansion. The first is *pixel replication*. When 2K pixel replication is chosen, each existing pixel in a row is copied three times, creating three pseudopixels. The first pixel is copied three times, and these three pseudopixels are placed, one after the other, next to the original pixel. Then three more pseudopixels are created from the second *original* pixel in that row of pixels, and placed in a line next to the second original pixel, etc. Eventually, the original row of 512 pixels is 2048 pixels long. This is done for each row of pixels in the original image, as well as for each pixel in each column. If a column contained 482 pixels, it would be expanded to 1928 pixels in a 2K pixel replication. The original file of 512 x 482 would become 2048 x 1928.

The second choice you are given is *pixel interpolation*. This is a very different method from pixel replication. The computer determines the color of the first and second pixels in a row, and calculates the three colors necessary to make a smooth transition between these two. For example, if the first pixel is a yellow-green, and the second pixel is a blue, the computer would calculate the colors needed for the three pseudopixels to make a smooth transition from the first yellow-green pixel to the second blue pixel, and then place these three pseudopixels in the row between the original first and original second pixels.

With pixel interpolation, the pseudodetail is real. (Admittedly a contradictory statement.) The pseudodetail is *new logical information* in the image. If part of a picture contains a leaf, with two adjoining pixels being slightly different greens, and one or more *new* interpolated pixels are created by the computer and placed between the two original green pixels, with each new pixel helping to make a smooth transition from the first original green pixel to the second original green pixel, then the new pixels make sense, both to the integrity of the image and to the eye.

Both of these methods work extremely well. When using pixel replication, you are only creating a larger file, in much the same way that you create a larger image when making a traditional photographic enlargement. When using pixel interpolation methods, you are in effect creating false detail. The pseudodetail may not be the same detail that was in the original scene, but it's just about impossible to know whether the detail in a pixel interpolated image was actually in the original image or not. Most people working with digital photography tend to use the pixel interpolation method, since it gives a smoother-looking image than pixel replication. We will discuss both of these methods again in Chapter 8.

Although HI RES is powerful, it is unnecessary for most digital photography purposes. If you have HI RES, you'll use it. At first, a lot, but as time goes on, less frequently. Soon you'll discover that working on a large file in many sections presents problems, and that a 512 x 482 (or fewer) pixel image is not only easier to work with, but will yield the same level of quality as a large original file, once it is interpolated to a high-resolution file (discussed in Chapter 8). You may find that you think of HI RES as you do a high-quality lens, which you only find occasional use for. Regardless, HI RES will allow a TARGA or TARGA-compatible to manipulate image files as large as those that Vista boards can manipulate and its subsampling feature is a great improvement over the *panning* (mentioned in Chapter 4) necessary to manipulate a Vista file with TIPS.

There is no denying that LUMENA, with or without HI RES, is a powerful program. But there is a price to pay for all its power. The list price of LUMENA for a TARGA-16 or 32, which now contains the Scan-In and HI RES modules, is $2,500. LUMENA for the Vista 4M costs $4,000. LUMENA works with the Vision-16 graphics board, as well as with the Artist RealVision-16 graphics board, although special versions for each board are needed.

Unlike TIPS which works well with a mouse, LUMENA was designed for use with a tablet. The program tends to rebel when it is used with a serial mouse, and may also do so when used with a bus type mouse. Time Arts strongly urges that you use a tablet with LUMENA, and I concur. This can raise the cost of LUMENA for a TARGA-16, with a tablet, to $3,000.

The manual that accompanies LUMENA is good. Because LUMENA can do many things TIPS can't, LUMENA is a bit more difficult to learn, since there are more combinations of tools and more optional settings available. If you start at the beginning of the manual, and try the tools described as you read about them in a methodical manner, you will soon become familiar with the logic behind the program.

LUMENA is copy protected. This is to protect Time Arts from having unauthorized copies of LUMENA floating around. They accomplish this copy protection by using a physical device, called a *copy block.* The copy block is a small plastic gadget that fits between your serial port and the cable from your tablet. Without this copy block, many of the functions of LUMENA will not work. You can make as many copies of the LUMENA program as you wish, because without the copy block they are not of much use. In essence, the copy block is worth $2,500. If you lose it, you'll have to negotiate with LUMENA for a replacement. It may help if you have filed a police lost or stolen property report.

ColorScheme II

ColorScheme II is a scaled-down version of LUMENA. It is only available for the Vision-16 board, and may well be the best reason for purchasing the Vision-16 over the TARGA-16. It differs from LUMENA in a number of important ways. First, it does not have all the tools found in LUMENA, but it does have many that you will find valuable. With ColorScheme II you will be able to "sandwich" two or more images together; correct perspective; merge part of one image with another (in a manner similar to, but more powerful than when using .win files in TIPS); create distorted images; sharpen, lighten or darken, and adjust contrast; emboss, and posterize.

ColorScheme II also differs from LUMENA in that it does not have a provision for using HI RES, nor does it have modules for running scanners or film output devices (discussed in the next chapter). None of these are major drawbacks. HI RES is fun to play with, but as mentioned earlier, many people won't make extensive use of it. Additionally, drivers are available from third-party manufacturers to control scanners and other input/output devices. ColorScheme II can be used with a mouse, it's not copy protected, and lists for $795. Time Arts has announced it will be offering "coupons" that can be used to upgrade from either ColorScheme I or II to LUMENA at a reduced price.

RIO

RIO (which stands for *R*esolution *I*ndependent *O*bject-oriented) is a program with limited value to digital photography, although it may be of more interest to those interested in computer-assisted drawing. The drawing tools found in RIO are not particularly suitable for digital photography, not because they aren't good (they are), but because they work in an incompatible system. TIPS, and most of the features of LUMENA, work on *bit mapped* (also referred to as *raster*) principles. This means that the computer keeps track of the image by remembering the color of each pixel in the image. If you change the color of a model's eyes, the computer recalculates the new colors, and changes the pixels at the indicated location. In effect, the computer says, "Change the 51st through 68th pixels in the 27th row from blue to green," and then does. RIO works on a *vector* principle, which creates mathematic equations for different parts of the image. This is fine if the image is made up of triangles, rectangles, circles, ellipses, etc. If you create a circle at a particular location of the image, the computer creates a formula for the size of the circle, and for its color and location. If you want to make the circle larger and move it to a new location, the computer makes appropriate changes to the mathematic formulae. This method has both advantages and disadvantages when compared to *raster* programs. However, vector programs are not useful for digital photography, because photographs contain too many irregular shapes.

Although RIO is inappropriate for digital photography by itself, in conjunction with TIPS it may be useful. RIO, as LUMENA, allows you to work on large-scanned files or to create large files from standard .tga files. Unlike LUMENA which can manipulate/create images of up to 2K resolution, RIO can manipulate/

create images up to 4K resolution. RIO, similarly to HI RES, uses subsampling (called *proxy editing* in RIO) to show the entire image on the screen at one time, and then allows you to choose part of the image to work on. After you select the portion of the image, you use TIPS to make modifications, then use RIO to replace (paste) the portion back into the entire image. However, all the problems discussed earlier in this book concerning working with large files are still an issue.

Another important feature found in RIO is drivers to run film recorders, printers (both discussed in the next chapter), and scanners. If you start with the incomplete version of TIPS supplied with your TARGA board and feel comfortable with it, and then decide to buy the complete version (or if you buy FreeStyle for your Vision board), you can purchase RIO and use its capabilities for handling large files (when you feel it's necessary) and for controlling input/output devices. The cost of RIO for a TARGA-16 is about $1800, and when added to the $795 for TIPS, the price is about the same as for LUMENA for the TARGA-16 ($2500), sans the needed tablet. Your TIPS + RIO system lacks some of LUMENA's features, but it provides drivers you may need to control peripherals.

UTILITY PROGRAMS

Utility programs encompass a broad range of programs that either add features to an existing program, make an existing program easier to use, or solve incompatibility or other types of problems. As I have indicated, LUMENA has a number of effects and features that TIPS lacks. One utility program, Ron Scott's *QFX*, is a special effects program that enhances TIPS, adding a number of features found in LUMENA. Among these features are the ability to sharpen a video or scanned image; the ability to adjust the contrast and brightness of the image; and to create posterizations, high-contrast, and solarized effects, as well as embossed and shadow effects (e.g., cast shadows in various colors from portions of the image). Many of these effects are achieved in ways other than those used by LUMENA, so the resulting effects may look different. It is a matter of opinion whether the effects of one program are better than those of the other. The cost of Ron Scott's *QFX* is $500, it's not copy protected, and the same version will work with TIPS with a TARGA-16, 24, or 32 board. As is the case with TIPS, LUMENA, and RIO, this utility program is slick and well designed.

Another general-purpose utility program is *Norman Breslow's Photo Utilities*, which is really a grouping of about 20 different programs. Unlike the programs previously mentioned in this chapter, these are no-frills programs designed primarily for ease of use. These programs were created to fill some gaps in the image making capabilities found in other programs, but they may duplicate some functions found in other software. These utilities can lighten or darken an image, change a color image to black and white, reverse colors into their complements, adjust contrast, and compensate for or create color shifts by allowing you to adjust the red, green, and blue components in the image. Additionally, they can make numerous different types of posterizations, either in color or black and white, make line resolutions (pen and ink effects) and high-contrast images, and create thousands of variations from one original image. A number of the programs allow you to make color images from black and whites, one of which gives a brown tone effect, and another a color-inferred effect. Another program makes black and white images from a color original using the equivalent of a red, blue, or green filter. This allows you to darken or lighten skies for dramatic B&W effects, darken or lighten skin tones, etc. When these B&W filter effects are used at their extreme settings, they tend to create charcoal drawing-like images. The cost of Norman Breslow's Photo Utilities is $65, they are not copy protected, and they are designed to work with TARGA-16 boards with files created by TIPS and LUMENA.

Following are descriptions of some other utility programs.

ASAVision, from CYNC Corp, is designed primarily for refining and manipulating video image input. Its most interesting feature is its "linear chromakey function," which is similar to chromakey (blue background) used for special effects in TV and movies. A model can be taped (or a live video image used) on a blue background and then superimposed onto an existing digitized image. This process is similar to the TIPS .win function (and similar tools in LUMENA), and to other TIPS or LUMENA video grab functions. It differs in that shadows cast by the model will also appear on the new combined image, giving a very realistic appearance. Furthermore, translucent objects, such as a glass of water shot on the blue background, will allow the background the image is superimposed over to show through. Other functions of ASAVision are a bend function, which will cause a wavy, fun-house mirror effect, posterizations, and image enhancement features. Its list price is about $300.

INPrint, available from Mathamatica, enables TARGA images to be easily printed on most laser printers. This is handy for creating a filing system of your images. You can print your images, write the file name on the paper, and file it away in a loose-leaf notebook. Although the laser print image is in black and white, and not of the highest quality, it is fine for cataloging purposes. The list price of INPrint is about $100. (A laser printer runs from $1,000 to $2,000 and up.)

MAC2TGA from Western Systems converts images from a Macintosh or Macintosh II file to a TARGA file, which can then be used with TIPS, LUMENA, and other paint and utility programs that can use TARGA image files. It lists for about $300.

Scan-T from Videotex Systems converts images from any black and white scanner into TARGA format. It lists for about $150. *Image Transformer* converts standard PC graphics (e.g., CGA or EGA) to TARGA format. It lists for about $200. *T-Scan* converts TARGA and Vision images to files used for desktop publishing, and lists for about $150.

These are examples of the many different types of utility programs that are available. It is doubtful whether anyone would need many of them, since for the most part they are highly specialized and meant to do only a few specific things. However, it's important for you to know what is available, since it is inevitable that sooner or later, one or more of these programs will come in handy. Knowing that it is possible to change a Macintosh image file to a TARGA file may make the difference between accepting an assignment or turning it down.

SUMMARY

In this chapter I introduced you to two devices and a number of programs used in making digital photographs. As has been the case in previous chapters, you will have to choose which options to select on the basis of your particular needs.

If you do little or no drawing, I suggest that you use a mouse. But if you want to incorporate drawings with your photography, and/or you decide to use LUMENA as your paint program, you will need a tablet.

Happily, all the major paint programs mentioned (e.g., TIPS, LUMENA, ColorScheme II, and RIO) are excellent, although RIO's effects are limited in their usefulness to digital photography. TIPS is the easiest to use and is also quite powerful. It will give you great control over your photographic image compared to the control traditional photography presently gives you. It is also inexpensive, as far as professional programs go. TIPS does have a number of deficits, however. It does not allow you to manipulate images larger than 512 x 482 pixels, so using a scanner to make large (high resolution) files is futile.

If you decide that you need to work with large-scanned files, or large files containing pseudopixels, then RIO can be purchased. The drawing functions of RIO will be of little value in digital photography, but it does give you the ability to either create large files or to manipulate large-scanned files. It also includes programs to run many peripheral devices.

A different approach is to get ColorScheme II, which is available only for the Vision-16 board. ColorScheme II is a close cousin to LUMENA, and contains those functions most important to digital photographers: the ability to change perspective, to sharpen, move, posterize, lighten/darken, etc., and all for about the cost of TIPS.

If your budget will allow, you should seriously consider the Rolls Royce of paint programs, LUMENA. It has all the tools a digital photographer needs and is *modular* so drivers and other functions can be bought from Time Arts and added to LUMENA's base system as needs arise.

I find that working with both LUMENA and TIPS as well as Ron Scott's QFX and Norman Breslow's Photo Utilities gives me the flexibility and the effects I want. One program may offer something the others don't, or may allow a simpler solution to problems than the others provide. My setup may be inappropriate for you but you should be aware that many programs can be used in conjunction with each other.

CHAPTER 8

THE FINAL IMAGE (DARKROOM)

Now that we have some understanding of what is required to get a photographic image into a computer, and what can be done with the image once it has been digitized, we must turn our attention to getting the image out of the computer and onto either paper or film.

In this chapter I will present five methods that are used to turn the digital image in a computer into either prints, negatives, slides, or transparencies (up to 8 x 10 inches). First I will describe shooting an image off your computer graphics monitor, then screen cameras, and film recorders, followed by a discussion of the relationship of file size, photographic quality, and film recorders. To conclude the discussion I present the pros and cons of paper printers and service bureaus. Finally, a brief summary of the information covered in this chapter will be provided, along with my recommendations as to which methods you should consider.

PHOTOGRAPHING THE GRAPHICS MONITOR

To take a picture of an image on your monitor screen, all you will need is a 35mm camera which allows for manual exposure control, a sturdy tripod, a way of darkening the room your computer is in, and some time and patience. This method involves placing your camera on a tripod, focusing on the monitor, making an exposure, processing the film, and making a print. While this is rather straightforward, it is rife with technical problems you should be aware of. Although the least expensive method for turning a digital image into a print, it unfortunately results in the worst quality. Nevertheless, this method may be of value to those of you just beginning digital photography, and therefore a lengthy discussion about it will follow.

When you set up your camera and tripod and focus on the computer monitor, you must make absolutely certain that the camera is parallel with the monitor screen. That is, an imaginary straight line from the center of the lens to the center of the screen must make a right angle with both the lens and the monitor screen. This cannot be emphasized too strongly. If you are not "squared up," the entire image will be distorted: One side of the image will be larger than the other.

Even if you do shoot at parallel to the monitor, you will still have some *pincushion distortion* in your print. Pincushion distortion causes lines to look as if they are slightly curved. This is because the monitor screen is slightly curved. The image at the center of the screen will be undistorted, but as you go further to the edges, you will start to notice distortion. The only way to avoid this problem is to use a flat-screen color monitor. Unfortunately, as of this writing, only one flat-screen color monitor is readily available, and it is not compatible with the TARGA and TARGA-compatible systems, but rather with VGA graphics. Flat-screen color monitors are not popular because they cost more to manufacture than the standard slightly curved monitors, and because they are not as sharp as standard monitors. It is more difficult to get the electron beam to accurately focus at the edges of a flat screen than at the edges of a slightly rounded screen. Because you would be spending more and getting less, not many manufacturers feel there is a large enough market for them.

For most photographic applications, this pincushion distortion will not be a major hindrance, because it will be most noticeable only at the extreme edges of the image. Because the computer monitor is closer in proportions to a 4 x 5 inch negative format than the 35mm format I recommend you use, the extreme sides of the digital image are usually cropped off, thus eliminating the most distracting areas of distortion. Because the top and bottom of the image will also be slightly distorted, you may find yourself cropping into this part of the negative when you make your print, again eliminating the major areas of distortion.

Even if you do shoot straight on to the face of the monitor, and even if you eliminate the most annoying areas of pincushion distortion, slight distortion will still be evident. However, if the image is of a landscape, or even a person, the distortion will be so slight as to be unnoticeable. If the image contains straight horizontal and vertical lines, such as found in architectural photography, you will find the distortion very noticeable and, usually, unpleasant.

It would be helpful if you could shoot just the central area of the monitor, thus getting a full-frame 35mm negative that did not need cropping to delete the curved edges caused by the slightly curved monitor screen. However, if you do this, the pixels will become distinct, and your resulting image will look as if it were made up of numerous dots. This is because you will be enlarging the pixels. This effect may work well for some purposes (some effects in fine art or high fashion photography, for example), but for most applications the resulting image will be unpleasant. Additionally, you will still have some distortion at the edges of the image.

What you need to do to avoid getting an image that looks as if it is made up of dots is to shoot an image that shows the inner edges of the computer monitor case (where the glass of the screen meets the plastic of the monitor cabinet). When you look at the image through the viewfinder of your camera, you may be able to see distinctly the pixels that form the digital image. This view is slightly magnified from that that will be on the film. The pixels on the film will be so close together that they will tend to "merge" together. This is why I suggested shooting the entire screen, including a small portion of the monitor cabinet. This image area will be sufficient to cause the pixels to become unnoticeable when small prints (about 3 x 2 inches) are made. If you make a print much larger than this, the image will deteriorate rapidly, looking grainy and showing the pixels that make up the image. And if you try to shoot the image on a full frame of 2 1/4 inch or 4 x 5 inch film, the pixels will also be evident. Only an image area smaller than a full 35mm negative will work.

Another difficult technical problem is that of getting the correct exposure and contrast when shooting off the monitor. The contrast of the image you see on the graphics monitor when you are working on the imagery may look good to your eye, but may be too contrasty (the usual problem) or too flat to yield an image with good photographic detail. In order to determine the correct setting of your monitor for your film you will have to make a number of test exposures using the following guidelines.

1. Your graphics monitor will have both brightness and contrast controls. These controls will differ in physical appearance depending on the type of graphics monitor you buy. Figure 8.1 shows the brightness and contrast controls on an Electrohome 1311 monitor. Regardless of the type of controls on your graphics monitor, you will have to devise a method of creating a scale that will allow you to (roughly) divide both controls into five sections. By this I mean that you should be able to adjust the brightness and contrast in quarterly increments from maximum to minimum. This means you will have maximum, -1/4, -1/2, -3/4, and minimum settings marked for both the contrast and brightness controls.

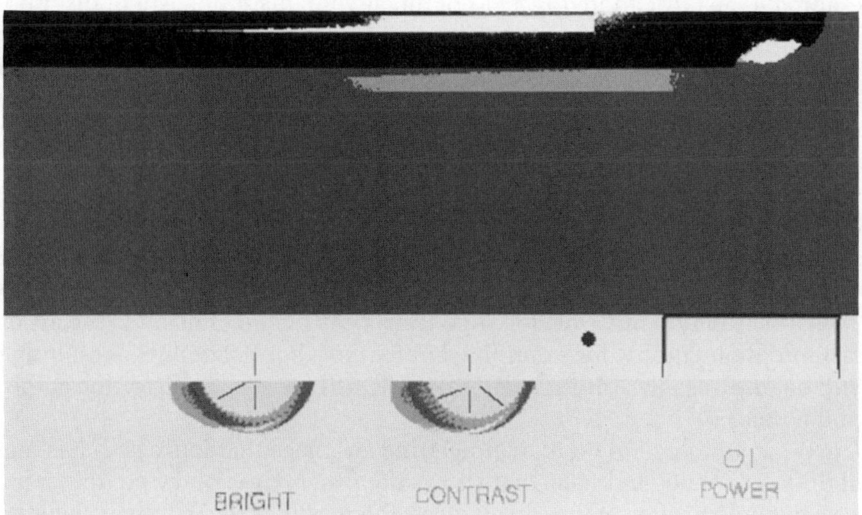

Figure 8.1 The contrast and brightness controls of Electrohome 1311 graphics monitor, using TIPS and utility programs.

It is all right if these increments are approximate. They should be marked with a pencil or felt-tipped pen on the monitor cabinet, as seen in Figure 8.1.

2. Using your paint program, make a gray scale from black to white to use as a test target. This can be one of the learning projects you use to get used to your paint program. Save this gray scale as a file. You may include some colors in the gray scale, along with some vertical and horizontal lines to help in determining if your camera is aimed straight at the monitor screen. Figure 4.2 shows such a gray scale.

3. Set your camera speed at 1/2 or 1/4 of a second. You must use shutter speeds slower than 1/30 of a second, since that is the time it takes for a complete image (frame) to be displayed on the screen. If you shoot at 1/30 of a second, you may find that part of the image is slightly darker or lighter than the rest of the image. This can occur if your shutter speed is slightly off. If your shutter is set at 1/30 of a second, but is actually shooting at 1/38 of a second, you will not get a full frame, but one field and most of the second field. This will cause some loss of image. If your shutter speed is slower than 1/30 of a second, you may find that you do get both fields (one frame), but part of the next field is also exposed, causing the equivalent of a partial double exposure, and thus a slightly lighter image on part of your print. If you shoot at 1/2 of a second, you will be shooting 15 complete frames. If it turns out that you shoot 15 frames and part of the next field, or slightly less than 15 full frames, the *very slight* over or underexposure will not be noticeable.

4. Because you are making a test, the film you choose to use at this time should be the film you will use when making photographs from your monitor. You can't make a test with one brand (and speed) of film and then shoot the images with another. I have found that Kodacolor Gold 400 or 200 will yield a *slightly* less contrasty image than Varicolor 400 or 200, but both will yield acceptable results.

Putting these steps together: Using your paint program, bring the file containing the gray scale with the vertical and horizontal lines onto your graphics monitor. Darken the room. Focus on the lines, and use them to try to determine whether you are shooting at a slight angle to the monitor. If you see the vertical or horizontal lines converging, reposition your camera/tripod until they look parallel. When you are "squared up," set the contrast and brightness controls of the monitor at maximum. Set your shutter speed at 1/2 of a second. Use your camera's light meter to get an exposure reading. Adjust your f/stop so that it is 1/3 larger (letting in more light) than the light meter indicates. Make an exposure, and write down the exposure time and the contrast and brightness settings of the monitor. Now close down the lens 1/3 f/stop to the normal setting indicated by your light meter, and make another exposure. Write down this information. Close down the lens another 1/3 f/stop, expose, and indicate the exposure and monitor setting information. You will now have three exposures: one overexposed, one normal, and one underexposed.

Keep the brightness setting at maximum and lower the contrast control on the monitor to the 1/4 from maximum setting. Take another light meter reading. This reading may be the same, or very close to the first reading. Now follow the above directions to bracket your exposures in 1/3 f/stop increments from +1/3 to -1/3 normal. Make sure that you keep accurate records of your exposure information for each frame you shoot.

Repeat this procedure for each contrast and brightness combination. In total, you should have 75 exposures to make. However, by the end of a 36-exposure roll of film, you will have made 36 different bracketed exposures for about half of all the possible brightness and contrast settings. This should be good enough. Develop the film and make a color contact sheet (or have the film processed and printed for you). Examine the gray scale. Choose the exposure that yields a pure black, a pure white, and shows all the grays in between. If your best print is too contrasty, that is, if you not only have a pure white and black, but some of the grays are also either white or black, then rerun the test, this time using the monitor settings that were not used on the first test because of lack of film. It has been my experience that one of the exposures on the first test roll should be adequate. You may want to do some fine tuning if one of the negatives is close, but not perfect. Try shooting some more negatives at the contrast and brightness settings used for that negative, but vary your f/stop somewhat to zero in on the perfect setting.

The advantages of this system are that it costs nothing above the cost of film, paper, and chemicals (or lab fees) and also, once you are zeroed in, you will be able to make perfect negatives and prints with little or no problems. This is due to the fact that the color characteristics of the color monitor will be constant, as will the exposure.

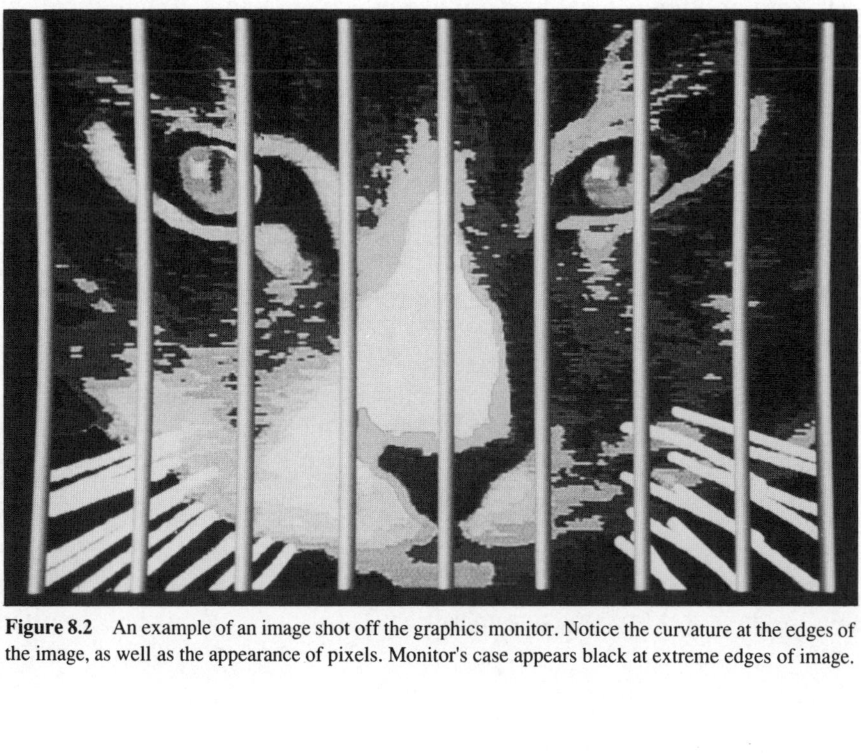

Figure 8.2 An example of an image shot off the graphics monitor. Notice the curvature at the edges of the image, as well as the appearance of pixels. Monitor's case appears black at extreme edges of image.

The disadvantage of this system is that some distortion in horizontal and vertical lines will be evident, and you will only be able to make small prints. Figure 8.2 is an example of what a typical color image looks like when shot off the graphics monitor. It was photographed off an Electrohome 1311 graphics monitor with Kodacolor 400 Gold film, using the procedures described above. This print is reproduced approximately 5 x 3.5 inches, and is of the entire negative. You can see individual pixels which make up the image on the monitor, as well as the distortion of the vertical and (to a lesser extent) horizontal lines at the edges of the image. If the image were cropped so that the monitor's case and major areas of distortion were eliminated, and still printed 5 x 3.5 inches, the pixels would be emphasized to a greater extent than they are now. By making the print *smaller* and cropping off the extreme edges of the image, the pixels become less noticeable or disappear completely, and the pincushion distortion is either completely eliminated or greatly minimized.

SCREEN CAMERAS

Although screen cameras are no longer readily available, they are described because they represent an earlier stage of technologic development which leads to the film recorders presented in the next section. Screen cameras consist of a color monitor enclosed inside a cabinet that also contains a camera with a close-up lens. The back of the camera protrudes from one end of the cabinet, and a cable attached to the computer graphics board plugs into the other end of the cabinet. The image is sent via the cable to a high-quality color monitor inside the cabinet where the exposures are made. Screen cameras resolve the problems encountered in the previous section, of camera/monitor alignment, brightness, contrast, and exposure, but all the other problems of shooting an image directly from a monitor screen still exist (e.g., distortion and discernible pixels). Screen cameras are priced at a few thousand dollars, although they are difficult to find today because they do not produce particularly good results compared to other methods available.

FILM RECORDERS

Film recorders represent the state of the art for obtaining quality negatives, slides, and transparencies up to 8 x 10 inches from computer-generated images. All film recorders have a superficial resemblance to screen cameras, since they too consist of a monitor enclosed inside a cabinet, with a camera back protruding at one end and a cable attached to the other. But here the resemblance ends.

Some film recorders only make 35mm slides, but others can make 4 x 5 inch and 8 x 10 inch transparencies, and some can make exposures on 16mm and 35mm motion picture film. To accommodate different film sizes, different cameras are inserted into the film recorder. Each camera has a lens designed to enlarge or reduce the image on the monitor to match the film size being used.

The monitor in a film recorder is not a color monitor, but instead a *flat-screen, black and white* monitor. A tricolor method of exposing the film is used. Information from the color digital image is broken down into its red, green, and blue components. First, information about the reds contained in the image is sent to the flat monochrome monitor screen, which creates an image made up only of this information. The image is made up of grays, with the intensity (brightness or darkness) of the gray representing the amount of the red in the digital photograph. For example, areas with a great deal of red will be very bright, and areas that contain little or no red will be dark. A red filter is automatically positioned by the film recorder in front of the camera's lens, and an exposure is made. After the red information has been recorded onto the film in the camera, the green information is displayed, with a green filter automatically positioned between the monitor and the camera's lens, and then this process is repeated for the blue information. The film does not advance between exposures; rather, a triple exposure is made on one frame of the film.

It should be noted that the full red, green, or blue image is not displayed on the black and white monitor at one time. Instead, each pixel in the first row is displayed, one at a time, then the pixels in the next row, etc. Depending on the complexity (e.g., file size and number of colors) of the digital photograph, the type of film recorder used, and the driver used to control the film recorder, this process can take from two minutes to a half hour or longer.

A *screen camera* takes information from the digital photography graphics board and sends it directly to the monitor in the cabinet in the same way it sends information to the graphics monitor you view the image on. The quality of the image is limited to the amount of pixels the monitor can simultaneously display. An image from a Vista/NuVista board will be superior to an image from a TARGA-type board because the latter displays fewer pixels on a monitor than the former. With more pixels displayed, more detail appears in the image. Additionally, with more pixels, problems with jaggies are reduced. A film recorder does not take information from the digital photography graphics board, but rather from the file containing the information about the image.

As we discussed in earlier chapters, it is possible for the file to contain more information than can be displayed on a monitor at one time. All this information can be displayed on the film recorder's monitor. This is accomplished by displaying one row of information at a time on the monitor. Some of the problems inherent in displaying a large amount of pixels, with a large number of simultaneous colors, involves having enough memory (RAM) to hold the entire image and rapidly send it to the graphics monitor, as well as the mechanics of getting the monitor to rapidly position the millions of pixels contained in a high resolution file on the screen.

Because the film recorder sends only one row of information to the film recorder's monitor at a time and only in shades of gray, both the memory (RAM) requirements and the problem of rapidly displaying millions of pixels in many colors are reduced, if not eliminated. Additionally, the monitors used in film recorders are capable of displaying up to 16K pixels per row, although most affordable film recorders can display only 2K or 4K pixels per row. The filter system allows the film to reconstruct the full color image, in much the same way color separation negatives/positives are used to make dye transfer photographic prints or color printing plates for lithography.

As with scanners and other *peripheral* devices, the film recorder is connected to the computer via a cable. The cable usually connects either to a special expansion board that is included with the film recorder, or to a GPIB. The film recorder contains RAM specifically used by the film recorder and that is not available for the computer's general memory needs. Even though information is sent one row at a time in shades of gray to the film recorder's monitor, larger files created with LUMENA or RIO may require additional RAM to be added to the RAM that comes with the film recorder.

To operate a film recorder, you run a program (driver), which is similar, in type, to the programs that run scanners. A series of questions will be printed on the standard monitor screen by the program and you will select the appropriate answers. After the questions have been answered, the program takes care of the rest.

The manufacturers of film recorders like to advertise that a color image can be photographed in a minute or two, but these images are usually low-resolution (small file) graphs and charts containing a relatively few colors. For a full-color 2K resolution file, the time may range from a few minutes to 20 to 30 minutes, depending on the design of the film recorder and the amount of RAM it contains. If you have ten images to deliver, you may have to build in a few hours of "shooting" time.

Many film recorders state in their advertising that they can work in *background mode*, which means that you can continue to use your computer for other tasks while a program runs the film recorder. This sounds good, but the same advertising usually omits mentioning that from 128KB to 256KB of your computer's conventional RAM may be necessary to operate the film recorder in background mode. Since programs like TIPS, LUMENA, and RIO need almost all the conventional RAM your computer has, you may not be able to use these programs and run the film recorder at the same time. However, you may be able to use your computer for other purposes, such as writing letters.

One way to handle long printing times when you can't use the film recorder in background mode is to use a *batch file*. Batch files are special DOS files you can create that allow you to give a list of instructions to the computer. In essence, the computer thinks the batch file is you typing instructions. This means that you can write a batch file to automatically run the film recorder at night, or while you are out of the studio. This frees your time to do other things. The film recorder must have an automatic film advance to enable this operation, and most do. When you create this file, you give it the file extension .BAT (briefly mentioned in Chapter 2). Information for creating batch files can be found in your DOS manual which comes with DOS, from basic books on using IBM-type computers, from a course on basic computing, and possibly with the manual for the program used to operate the film recorder.

When film recorders were first introduced, many had problems with *raster scan lines*. These are the scan lines that make up the image on a television set or monitor screen. If you look closely at a TV set, you will notice that there is some space between each line of information. (Some television sets hide these lines with an intricate masking system, which just disguises the lines, but does not do away with them.) These raster scan lines appeared on the film. Methods of *raster fill* were developed that eliminated the space between the lines. Today, almost all film recorders being sold have successfully eliminated the presence of raster lines. If you should buy a film recorder, inquire whether the system uses raster fill and ask to see sample slides made from the recorder.

Another consideration which will directly affect the outcome of the finished image is the number of colors film recorders can produce. Some can only simultaneously display 256 colors, even if the original digital image is made up of 32K or 16.7M colors. Other film recorders can print 16.7M colors, and some even more. The different number of simultaneously displayable colors has nothing to do with the cost of the film recorder. Some expensive ones can only output a relatively few colors. This is no handicap if you are making graphs and charts, but it will limit the image quality if you are making full color photographs.

A final consideration we should note is whether the film recorder is *analog*, *digital* or both analog and digital. Analog film recorders are very similar to screen cameras in that they use an image sent by the computer's digital photography graphics board, but differ from screen cameras in that the image is sent to a flat-screen gray-scale monitor. As is the case with screen cameras, the maximum resolution of the image the camera shoots is limited to the maximum number of pixels the monitor can display at one time, which will be far less than the number of pixels needed to create a jaggie free image. However, analog film recorders will work with any program that can display an image on a monitor's screen, since they do not use a driver and image file. So you get a trade-off between lower quality and the ability to capture any image made by any program using any (i.e., EGA, VGA, TARGA, etc.) graphics board.

Digital film recorders use information contained in files and are dependent on a driver to act as an intermediary between the program that created the file and the film recorder. These film recorders give the highest quality, but will only work with specific programs. If the program you create your digital image with does not have a driver for the film recorder you want to use, and if no third-party manufacturers make a driver to allow the program and film recorder to work with each other, then the film recorder and the paint program are incompatible.

Finally, some film recorders are capable of working in either analog or digital mode, but these are rare, and one could argue that a film recorder that tries to be all things to all people is not as good as one that does one specific thing. It is fairly well accepted that for quality work, you want a digital film recorder. When purchasing one, you must make sure that it either comes with a driver that will work with your system, or that such drivers are available from third-party manufacturers.

Because film recorders vary so widely in technical specifications, such as the number of drivers available for them, and the amount of RAM necessary to run them in background mode, etc., it is best to discuss your needs with Truevision/Vision Technology or other specialized graphics resellers. Following are a number of film recorders that are representative of those on the market, to give you some idea of what is available.

The *Lasergraphics Film Recorder (LFR)* is capable of creating either 2K or 4K resolution images on 35mm slide or negative film with ISOs (light sensitivities) of 40, 100, or 200, or on 3 1/4 x 4 1/4 inch Polaroid film, in up to 16.7M colors. The 35mm camera (currently either a Nikon N-2000 or Pentax PT135) comes with automatic film advance. It comes with a program that makes it compatible with TARGA boards, and supports LUMENA and RIO. The LFR works in *background mode*, which means that you may continue working on images while the film recorder makes its exposures. Since it has automatic film advance, it can be run by a batch file. The LFR lists for just under $10,000. It comes with one of two different expansion boards. One, called the RASCOL II, is for IBMs and IBM-compatibles, and the other, called MacRASCOL, is for the Macintosh. For additional memory needed to work with high resolution files, add another $1,000.

The Lasergraphics PFR is a lower cost version of the LFR. LUMENA and RIO have drivers for it, and it comes with drivers for TIPS, LUMENA, and RIO. It lists for about $5,600.

The *Montage FR1* can create 2K or 4K high resolution images in 16.7M colors, on 35mm ISO 100 slide film. It comes with its own expansion board. The camera automatically advances (which means you can use batch files) and automatically rewinds after the last exposure. The film recorder is TARGA-compatible, and comes with drivers for ColorScheme I and II, FreeStyle-16, TIPS, LUMENA, and RIO, as well as many CAD programs. It has a 90-day warranty, and lists for about $6,000 (without additional memory that may be necessary to manage large files).

Matrix Computer Graphics, which is a division of Agfa, has been considered the leader in film recorder technology for personal and small business use, mainly because of superior image quality. Their *PCR* film recorder is capable of creating 35mm slides or negatives in 2K or 4K resolution, with up to 16.7M simultaneous colors. It uses a GPIB or one of two special Matrix expansion boards (which cost from about $2,000 to $2,500, but aren't needed for most digital photography purposes). It has automatic film advance so batch processing is possible. Drivers are available for it from LUMENA, RIO, and from other third-party software developers. It lists for about $13,000.

The Matrix *QCR-Z* creates 2K and 4K resolution 35mm, 4 x 5 inch or 8 x 10 inch images in up to 16.7M simultaneous colors, and works faster than many of the film recorders mentioned so far. As with the PCR, drivers from RIO, LUMENA, and other third-party software developers are available. The QCR-Z costs about $25,000.

Management Graphics makes the *Solitaire-8* which can record at 2K, 4K, and 8K resolutions, with up to 16.7M simultaneous colors, at very fast recording times. It creates images in 35mm to 8 x 10 inch format. The *Solitaire-16* is similar to the Solitaire-8, but has a top resolution of 16K, and a maximum of over 68 billion simultaneously displayed colors. The Solitaire-8 lists for about $35,000 and the Solitaire-16 for about $75,000.

PHOTOGRAPHIC QUALITY, FILE SIZE, AND FILM RECORDERS

At this point it is best if we conclude the discussions of high-quality, large versus small files, and jaggies that we have had, in small sections, over the past few chapters, and put them in context with film recorders.

File Resolutions and Film Recorder Resolution

The size of the file and the resolution the film recorder is working at must match. If you have 512 horizontal pixels per row in a TARGA or TARGA-compatible image, and reproduce it at 2K resolution on 35mm film, the resulting image will be much smaller than the image area of the film. If the film recorder is printing pixels close enough together to get 2048 pixels in a row on 1.5 inches of film, then 512 horizontal pixels would take up slightly more than 1/3 of an inch, leaving most of the film blank. To avoid this situation, files smaller than the resolution the film recorder is printing at are enlarged to the resolution being used by the film recorder.

In the previous chapter we discussed pixel interpolation and replication, and the ability of LUMENA's HI RES to enlarge a TARGA image to 2K resolution. If the image file is at 512 x 342 pixels (35mm ratio), then the interpolation/replication functions of LUMENA, RIO, or a program from Ron Scott called SQALE (list price about $400) can be used to interpolate the file to a higher resolution. After the image file has been expanded to match the resolution the film recorder is operating at, the image file is read by the film recorder, and the data are used to make the slide. It is not necessary to actually manipulate the expanded image. Hence, if you use TIPS at 512 x 342 pixels, and then expand the image with SQALE, you will have the file in the proper condition to use with a film recorder.

But manually scaling up the smaller image file is not necessary, because drivers for the film recorder will automatically do it for you. A driver is an *accommodating translator.* It is a translator because it acts as a "go-between" for a program and a peripheral device, such as a film recorder. And accommodating, because it will put the information in the file in a form needed by the film recorder. If you have enlarged the file to the proper resolution (e.g., the resolution the film recorder is working at), by using either LUMENA's HI RES, RIO, or Ron Scott's SQALE, or working from a large-scanned image, the driver won't alter the file. If the finished file is only 512 x 342, the driver will scale the file to the proper resolution automatically. For example, if a film recorder can work at either 2K or 4K resolution, and it is set to run at 2K resolution, the driver will enlarge the file, either by pixel *replication* or *interpolation* (depending on the driver) to 2K. If the film recorder is set to run at 4K, then the file will automatically be expanded to 4K size.

It can be argued that a large scanned image is superior to a smaller image that is later expanded, at least as far as true photographic detail is concerned. Technically speaking, this is true. But pixel interpolation creates false detail which is logical detail, and looks as if it were real detail. Some people will prefer working with *true* detail, and this preference will outweigh problems of working on an image in sections; others will opt for the relative ease of working with only one section (e.g., a 512 x 342 pixel file) and will use pseudodetail.

Film Size and Photographic Quality

As we have seen, some film recorders only make 35mm slides, and others can also make 4 x 5 inch or 8 x 10 inch images. In traditional photography, the larger the film size, the higher the quality of the image. In digital photography, this is not necessarily true, and often the opposite is the case.

The size of the film your finished image is put on will have a major effect on the final quality of your digital photograph. If you made an image at 2K resolution on 35mm film, you would have 2048 pixels in each row measuring 1.5 inches, which is equal to 1366 pixels per inch. If each pixel were a square and touched each of its adjoining pixels, then each pixel on 35mm film would be 1/1366 of an inch square. At 4K resolution, you would have 2732 pixels per inch, and each pixel would be 1/2732 of an inch square.

With 4 x 5 inch film, a 2K resolution image would contain 2048 pixels over 5 inches, thus the "pixel density" would be 410 pixels per inch. The size of each pixel would be estimated at 1/410 inch square in each inch. At 4K resolution, there would be 819 pixels per inch, and each pixel would be 1/819 of an inch square.

With these figures, we can see that at 2K or 4K resolution each pixel on a 4 x 5 inch transparency is about 3.3 times larger than a pixel at the same resolution on 35mm film. The larger pixel size makes the image look grainy or "blocky;" individual pixels begin to be discernible rather than a smooth transition from one pixel to the next. Photographically speaking, the image starts to fall apart. At 4K resolution on 8 x 10 inch film, the image starts to look like a mosaic.

Earlier in this book I mentioned that the more pixels you have to work with, the higher the resolution of the image. But if you have more pixels distributed over a larger physical space, you will not benefit from the increased number of pixels. What you need is more pixels in a smaller physical space. What counts is pixel density. High resolution is gained by having a large number of pixels per unit of area. The only way to achieve 35mm quality with larger film is to increase the number of pixels. To get the equivalent of 2K 35mm quality with 4 x 5 inch film, you will need to have an 8K file, and for 8 x 10 film, a 16K file will be necessary so as to maintain a high level of pixel density.

Film recorders capable of 8K resolution and that also use 4 x 5 or 8 x 10 inch film, cost at least $35,000. Because this is out of the price range of most individuals, you may have to limit your output to 35mm film. Since a 35mm image at 2K will yield a superior image to a 4 x 5 inch image at 4K, don't be seduced into buying a film recorder because it can make 4 x 5 or larger images, unless it can create these images of at least 8K resolution or higher.

Jaggies

I have mentioned jaggies from time to time, and pointed out that they are caused by having too few pixels to create a smooth shape, usually those containing curves. The only way to eliminate jaggies is to have files that contain enough pixels to allow smooth-looking lines. The important point is that it is the image in the file that counts, not the image on your digital photography graphics monitor. If you see jaggies on your graphics monitor at a resolution of 512 x 342, they will be greatly diminished at 2K resolution, and will probably disappear at 4K resolution. BUT ONLY IF THE FILE HAS BEEN PIXEL INTERPOLATED. If the file is enlarged by pixel replication, you will end up with *exactly* the same image, jaggies and all, at the higher replicated resolution. All you are doing with pixel replication is making an enlargement. With pixel interpolation, you are also making an enlargement, but the new pixels are not identical to the old ones. As mentioned earlier, these new pixels are logical, and will thus tend to make smoother transitions from one original pixel to the next. Jaggies represent a real weakness in the quality of digital photography. The only way to be certain to eliminate them completely is to work from large files. If you start with a 2K or 4K resolution scanned file, you will have less problems with jaggies than if you start with 512 x 482 (or fewer) resolution file and interpolate. Furthermore, interpolating the 2K or 4K scanned image to 8K or higher and using a film recorder that is capable of working at these high resolutions will assure that your final image is completely free of jaggies. The complete elimination of jaggies is the only reason to work from large files from scanned images, but the trade-off is that you will have to work on the image in many sections, and you will use a lot of disk space when making them. If you work with a 512 resolution file, and pixel interpolate the final image to 2K or 4K resolution, jaggies will be greatly reduced, if not completely eliminated. At present, jaggies are the weak link in digital photography. I will have a final point to make about jaggies in Chapter 11.

Some drivers only work by pixel replication, and others by pixel interpolation. For the most part, pixel replication should be avoided if you want the highest possible quality. If you have LUMENA and want to use pixel replication for a desired effect, fine. But if you send your images off to a service bureau, and they only use pixel replication, you will not get images free of jaggies. Additionally, if you buy a film recorder that comes with a driver for your graphics board/paint program, or if you buy a driver from a third party, and it only works by pixel replication, you will be dissatisfied with the resulting images, unless you work from very large scanned files.

File Ratio and Film Ratio

A final point which should be addressed is film ratios versus the ratio of the image on your computer graphics monitor. You should work on an image in the same proportion on your screen as the ratio of the film you are planing to use to output the image on. For example, 35mm film is in a 1.5:1 ratio, and 4 x 5 inch film is in a 5:4 ratio, but the ratio of the computer graphics monitor is 4:3. If you created a full-screen TARGA image (512 x 482 pixels), it would be in the wrong proportions to be easily used with 35mm or 4 x 5 inch film.

It is best to plan ahead. If you know you will be outputting to 35mm film, you should work with an image of 512 x 342 pixels, or some multiple of it, such as 1024 x 684 (1K resolution). To work from a 1K (or larger) file, you will need a program such as LUMENA with HI RES, or RIO, and then work in sections, since the TARGA (or similar) boards will not display images larger than 512 x 482. If you know you will be outputting to 4 x 5 inch film, you should work in a ratio of 512 x 384 (or some multiple of it, such as 1024 x 768, which is also called 1K resolution). You can set a scanner to give you these proportions.

But what if you have a previously scanned image at 512 x 482 pixels, or a video image that is in a 4:3 ratio and so not at the correct ratio for the film you will be outputting to? You can use a program such as SQALE, or functions in TIPS, LUMENA, or RIO to adjust a scanned or video image to the proper proportions. Unfortunately, doing this will cause distortions in your image. If you had a 512 x 482 resolution file, and you wanted to make it into a 512 x 342 resolution file, you would "squash" the 482 pixels per column into 342 pixels per column. A 5' 8" 120-pound fashion model would look like a 5' 4" 150-pound woman. If you want this type of distortion, fine, but if not, you will have a problem.

There are two ways to get an image that is not in the correct format to print onto film without distortion. They are called cropping and *fitting*. In essence, what you do is either leave some of the information at the top and bottom of the image off the final film image (cropping) or you can print the entire file onto a smaller area of the film, leaving some unexposed area surrounding the left and right side. In traditional photography, if you had a 35mm negative and wanted to print it onto 8 x 10 paper, you would either have to crop off some of the negative area, or print the entire negative onto the paper, leaving some of the paper surrounding the negative unexposed. Cropping or fitting are the equivalent in digital photography to the choices you have to make in traditional photography.

PRINTERS

Laser printers are the current "state-of-the-art" paper printers used for desktop publishing and for general office work. They are not particularly suitable for digital photography, since the black and white images they create usually have a maximum of 16 shades of gray and produce images of lower quality than those found in newspapers. However, it is possible to print TARGA images on them, and as mentioned in the previous chapter, a laser printer can be used to catalog your digital images.

Thermal-Transfer Printers (TTPs)

TTPs print a color image on plain paper. The process involves using multicolored "ribbons" made of wax impregnated with dyes. The ribbon is struck on one side by numerous heated pins pushed out by the printer, which melt the wax and cause it to stick to the paper. By melting different amounts of magenta, cyan, yellow, and black wax, a full color image can be formed. To create a continuous tone, color image effect, manufacturers use *dithering* patterns to try to create an optical illusion that blends the four colors together. As an example, for a light yellow, an attempt is made to surround the yellow dots with the white paper, but for a dark yellow, an attempt is made to surround the yellow dots with black dots. There are numerous different dithering schemes employed by different manufacturers, but none are particularly successful in their illusion.

After one line of dots has been printed, the paper is moved (usually 1/300 of an inch) and another line is printed. This resolution of 300 dpi is too low to form a high-resolution image; film recorders at 2K resolution using 35mm film have 1366 pixels per inch. The total time is usually two or three minutes for a 8.5 x 11 inch image to be printed.

A major problem with this technique is that all the dots are of the same size. If you look at a color reproduction in a magazine, you will notice that the color dots that make up the image are not uniform either in size or in space. They seem to be randomly distributed. (They aren't, but they look as if they are.) The evenness of both the spacing and the size of the dots formed by the pins in thermal-transfer prints causes an unrealistic image. There are no gradual tonal changes, even with dithering techniques.

There are two standard sizes of output from TTPs: *A size*, which is 8 1/2 x 11 inches, and *B size*, which is 11 x 17 inches, with equivalent (A4 and A3) European sizes available. Some TTPs are capable of printing only A size (or A4), while others can print both A (or A4) and B (or A3) size. It has been estimated that typical four-color prints (e.g., made out of magenta, cyan, yellow, and black) cost about 35 cents each in B size. Unlike many film recorders, TTPs are relatively fast, making a B size print in less than 5 minutes.

TTPs work in a number of differing ways. For some, a file is sent from your paint program via a cable to the TTP, which has its own microprocessor that takes care of dithering and other printing considerations. Some TTPs will need a pixel interpolated file, or large scanned file, to print an image the maximum size of the paper. As was the case with film recorders, drivers are necessary for this type of TTP. Other TTPs are similar to the screen cameras mentioned earlier in this chapter. These are analog TTPs, which take the image on the computer's graphics monitor and make a print directly from it, and not from a file.

If the TTP you are interested in uses files to create its image, check your paint program to see if the TTP is supported (i.e., if the paint program has a driver for it). If not, check with the manufacturer of the TTP to see if they have drivers that work with your paint program.

The cost of most TTPs is in the $4,000 to $10,000 range. Among the leading manufactures of TTPs are Mitsubishi, Seiko, Sharp, Shinko, Tektronix, and Versatec (to name just a few). A number of these companies are listed in the Sources section of this book, and resellers of digital photography equipment can help you decide which is best for your purposes.

The advantages of TTPs are that they are relatively fast, and for purposes such as industrial photography or package design, are relatively inexpensive. The disadvantages of TTPs are that they are relatively low-resolution devices that do not allow either for subtle shadings or images without noticeable dot patterns.

Ink Jet Printers

Ink jet printers work by spraying small dots of magenta, cyan, yellow, and black ink onto plain paper. The quality of the images generated by these printers is similar to that of images created by thermal-transfer printers. As with the

TTPs, most ink jet printers use a limited number of dots per inch, and with many ink jet printers, the dots are all of a uniform size.

Ink jet printers are good for posters and other forms of artwork, which may not need subtle shadings to make their point. They are also good for showing clients "roughs" of work in progress, and for communicating ideas within a company. The cost of a *Hewlett-Packard PrintJet* is about $1,400 (list) but it can often be bought at a discounted price. For this low price you get very low resolution (180 dpi) and a very limited palette (330 colors). The PrintJet plugs into one of the I/O ports in your computer. It can be used with TARGA images with a special driver available from Hewlett-Packard.

The *Sharp JX-730* color ink jet printer also has a resolution of 180 dpi, but it has variable dot sizes. Each of the four colors (magenta, cyan, yellow, and black) can be any of 12 different sizes. With dithering and variable dot sizes, images look better than those of ink jet printers that only use one size of dot. The JX 730 lists for about $2,200.

Although most ink jet printers have relatively low resolution and limited color ranges, the Iris ink jet printer, at $80,000–$120,000 offers superior color images. Iris achieves the increased quality mainly by varying the size of the dots it shoots onto the paper, to a far greater extent than the Sharp or similar types of ink jet printers do. The resulting image looks like a high-quality lithograph of about 160 lines. For comparison, a typical newspaper image may be in the 65 to 80 line screen range, while the typical color magazine reproduction is usually 120–133 lines. The Iris is also capable of printing on any surface up to 24 x 24 inches in size that is flexible enough to bend around the drum that holds the material in front of the spray jets. This means that high-quality photographic images can be put directly onto fabric, creating samples or one-of-a-kind materials.

Cycolor (Cylith) Paper Prints

Cycolor represents a dry photographic process, in which a traditional-looking photographic print can be made from a digital image, or directly from a slide. The process involves a photographic-like paper that has microencapsulated dyes (called cyliths) embedded in it. Light forms a latent image, equivalent to a latent image created in traditional film or photographic paper when it is exposed to light. However, this latent image is not formed by silver, but by the cyliths. Pressure is used to release dyes in the exposed cyliths, and transfer these dyes to paper or film, forming a color image. A brief heating process is used to help increase the color fidelity of the image. This technology is still new. The color is not yet as true as color found in traditional color photography, and the cylith image tends to be a bit contrasty. The richer colors (caused by the higher contrast) may be of value to some, but may be distracting to others. The process uses a special printer that is attached to your computer by a cable and is operated by a software program. Improvements in cylith technology are underway to bring the image up to traditional photographic standards.

Dry Silver Color Prints

Similar to the microencapsulated dye process used by Cycolor is the dry silver color process, which uses light to expose silver that in turn forms a latent image, similar to the way traditional color films and paper work. However, instead of using chemicals to release dyes and form an image, this technology uses heat at 270°F to release encapsulated dyes in the paper, forming the image. This process is being developed by Honeywell in conjunction with 3M.

Other photographic, dry print processes are available. Sony makes the UP-5000 printer, for example. The Sony process takes the video image directly from your graphics monitor and uses that information to create its image, which is similar to the way screen cameras and analog film recorders work. While the color and tonal range of the Sony prints are quite good, the overall resolution is that of your computer's monitor screen. The most notable visible problem is jaggies. Since this printer takes its image directly from a video image and not from a high-resolution file, it can make prints directly from a videotape camera or video and still video recorder, which may be of interest to some readers.

Thermal Dye Transfer Printers

Kodak has announced the XL-7700 Digital Continuous Tone Printer. Unlike the Sony UP-5000, it takes information from a computer file to form the image, and the process uses dyes that are transferred onto paper, and not encapsulated within the paper. I have seen samples of this process, and they look quite good. Unfortunately, as of this writing, the printer is not available for sale, although it is scheduled to be by the time this book is published. Although information about it is somewhat limited, it is known that the XL-7700 will make 8.5 x 11 inch and 11 inch square continuous tone color prints and transparencies. A less expensive gray-scale model will also be available. A full-color print takes about 3.5 minutes to make, and a gray-scale print takes about 1.5 minutes. An IBM model uses a GPIB, and a Macintosh model is also available. TARGA compatibility and price have not been announced. It can be assumed that the XL-7700 will be TARGA-compatible, since other equipment made by Kodak, such as their scanners, are compatible with TARGA boards (with optional drivers available from Kodak).

SERVICE BUREAUS

Just as service bureaus can scan an image to a disk file for you, they can be used to create from your paint file 35mm slides, 4 x 5 inch or 8 x 10 inch transparencies, and in some cases, negatives. The costs vary depending on the size of the file, the size of the image you want, and sometimes whether you need the image to be fitted or cropped, among other considerations. Full-color images from TIPS, RIO, or LUMENA programs run from a low of about $7 to over $30 each.

Some service bureaus specialize in particular areas. Jetgraphix, for instance, will scan images for you, but they don't put images onto film. Instead, they make large prints, either 24 x 30 inches or 32 x 40 inches, using an industrial grade ink jet printer. These images are full color, but when viewed close up they reveal noticeable dots. However, when viewed from the proper distance for such large prints, the dots disappear. Images such as these are good for posters and sales promotional graphics which are meant to be viewed at a distance.

There are two points to keep in mind when selecting and using a service bureau. First, the promised turnaround time varies from "same day service" to a week. Take any turnaround claims with a large grain of salt. Sometimes the service bureau gets backlogged with many orders, and other times their one very expensive film recorder may be out of order. It's best, if you're working on a tight deadline, to call and check in advance the status of the service bureau you are considering using.

Second, the prices charged by service bureaus vary greatly. One company charges $100 to *fit* a file, while another offers the same service free of charge. Because service bureaus are a relatively new phenomenon, one may wonder whether different service bureaus charge whatever the market will bear. This may be so to some extent. Another factor involves the type of equipment the service bureau is using. One company may charge more to make a slide from a file because their film recorder takes a half hour to do the job, while another company may charge less because their film recorder works faster, and so they can be more productive. Therefore, prices may reflect anything from price gouging to differences in equipment or the care the service bureau takes with your order. Moreover, a high price does not necessarily mean high quality, just as a low price does not necessarily mean low quality.

When you start reading computer magazines, you will notice some ads for service bureaus. Not all will work from TARGA-type files. Some only work from EGA/VGA files, used mainly for business graphs and charts. These companies usually state that their slides cost about $4. Before sending work off to any slide bureau, call and ask the following questions:

1. Do you make slides (prints) from TIPS (or compatible), RIO, or LUMENA files?

2. What information do you need me to send along with my file?

3. Do you take images by modem?

4. What is your turnaround time?

5. If I have a tight deadline, do you have a rush service?

6. At what resolution do you make 35mm slides? Only 2K or is 4K also available?

7. At what resolution do you make 4 x 5 inch or 8 x 10 inch transparencies? (If less than 8K, have 35mm slides made for the best quality.)

8. Can my files be pixel interpolated instead of replicated? (Interpolation adds *pseudopixels.*)

SUMMARY

The least expensive method of getting prints from a graphics file is to shoot the image off your graphics monitor screen. However, this is also the method that yields the lowest quality. Usable prints up to 3 x 2 inches can be made, but the quality of prints larger than this will rapidly fall apart.

Thermal Transfer Printers (TTPs) give quick prints in full color, but they lack the detail and modeling (gradual tonal changes) we are used to in traditional photography. They are good for posters, and may also be useful in industrial photography when it is necessary to show many variations, or to show work in progress to clients. Of course, these images may give you desirable effects for certain other jobs, and thus should not be dismissed out of hand.

For the highest quality at this time, you must use a film recorder. The finest quality presently available at affordable prices generate 35mm slides/negatives at 4K resolution; 4K 4 x 5 inch transparencies are not suitable, and a 35mm slide at 2K will give you a superior image. Transparencies that are 8K at 4 x 5 inches will yield slightly better results than 35mm at 2K resolution. If you want 8 x 10 inch transparencies, you will need 16K resolution for good results. However, as of this writing, I know of no service bureaus that use film recorders capable of making 8K or greater resolution images.

If you are planning to buy a film recorder, be certain to get samples of any you are interested in to check for the presence of raster scan lines, and to see the sharpness and general quality of the image it makes. It would be best to have a slide made, at 2K or 4K resolution, from one of your own images, because most of the sample slides provided by film recorder manufacturers are of charts and graphs, which won't give you a very good idea of how a high-quality continuous tone photograph will look.

Unfortunately, most film recorder manufacturers are reluctant, or even refuse, to make images from your files. This may be because they don't want to act as a service bureau. It takes time, and costs a few dollars, to make the slide. If they did this for everyone, the costs would mount. On the other hand, given the amount you are intending to invest in their product, you have a right to determine whether the recorder is capable of performing as you need it to.

By the early to mid-1990s, processes such as Kodak's XL-7700 paper printers, which yield full-scale, normal-contrast color and black and white images, will be available. At that time they may become the output device of choice for digital photography, since they will produce a full-color print in a few minutes and eliminate darkroom time and costs.

CHAPTER 9

THE SYSTEM: PUTTING IT ALL TOGETHER

In the previous chapters I discussed the base computer system, graphics boards, input and output devices, and programs used to manipulate the digitized image. In this chapter I will use the vocabulary introduced in those chapters to succinctly describe the process of digital photography, and then I will compare the cost of digital photography equipment to equivalent traditional photography equipment.

HOW THE PIECES FIT TOGETHER

By now you have a pretty good idea of how digital photography works. Your system consists of a base computer in which you have a digital photography graphics board. You input a photographic image using either a video camera or scanner, manipulate the image using a paint program, and then output the finished image by means of either a film recorder or a paper printer.

Your choice of a base computer is limited to those that are recommended for use with your digital photography graphics board. With the exception of the NuVista which was designed for the Macintosh computer, all of the boards discussed in this book require or suggest an IBM AT-type ("286") computer, with a bus speed of 8MHz or slower, and a BIOS that is fully compatible with the original IBMs (either Phoenix or Award will do).

Your choice of a digital photography graphics board is your single most important decision. If you plan to limit your work to B&W, both the TARGA-8M and the Vision-8 will give you 256 shades of gray, more than enough to do quality photography. Both of these boards are capable of displaying 256 simultaneous colors, but they are not recommended if you plan to do extensive color work. If you plan to do much work in color, the TARGA-16 or Vision-16 is recommended since they have a palette of 32K colors, and there are many paint and utility programs available for them. The TARGA-32 has marginally improved color, and will also give a 256 gray scale, but at a considerably higher cost than the TARGA-16 or Vision-16. Furthermore, there are fewer utility programs for the TARGA-32 than for the TARGA-16.

Using a video camera to input an image is simple, and may appeal to many photographers. However, the lack of quality, particularly the reduction in sharpness as well as highlight and shadow detail, will deter many from choosing this approach. This lack of quality is the result of the many steps required to create the digitized image. First the camera translates the incoming image to numbers, using a CCD or similar type of computer chip, then these numbers are changed into a video signal that is transmitted to the digital photography graphics board, where the video signal is transformed into numbers meaningful to the computer. It is this entire process, and especially the less than optimum quality of the digitizing feature of the digital photography graphics board, that causes the reduction in image quality.

For high-quality input, a scanner is recommended. The better quality of the scanned image is a result of the directness of the process involved: The print or slide is scanned with a CCD chip, and the resulting numbers are transferred directly to the computer where they are stored in a file. Gray-scale scanners will work well with the B&W graphics boards, although utility programs will probably be necessary, since gray-scale scanners are not designed for use with either the TARGA-M8 or Vision-8. More expensive color scanners are required for color work.

Although scanners enable the creation of high resolution files, there are two problems with using scanners on the highest resolution settings. First, the very large files they create take up a lot of room on your hard disk, and are difficult to transport from one locale to another. Second, these files are so large that they must be worked on in sections. At present there is no solution to the first problem that does not involve the renunciation of high-resolution files. The second problem may be avoided: Use a scanner to create small files of 512 resolution, manipulate the image with your paint program, and then use pixel interpolation to create pseudopixels immediately prior to outputting the final image. You can do this pixel interpolation manually, with the appropriate software, or you can let the driver for the film recorder expand the file for you. If you choose this option, make sure the driver uses pixel interpolation and not pixel replication.

Depending on your graphics board, you will have either a limited or a fairly wide range of paint programs to choose from to manipulate your image. With the TARGA-8M, you can well be limited to TIPS-8, although some programs designed for desktop publishing may be usable. For the Vision-8, your choice is limited to FreeStyle-8. For the TARGA-16, there is TIPS-16 and LUMENA-16, and for the TARGA-32, either TIPS-32 or LUMENA-32. For the Vision-16, there is ColorScheme II and LUMENA-16. For the TARGA-16 and 32 (and some compatible) boards, there is a host of utility programs that can be used to either add features to paint programs for these boards, enhance features that do exist, perform other functions, such as file conversions from one format to another, or drive equipment such as scanners and film recorders.

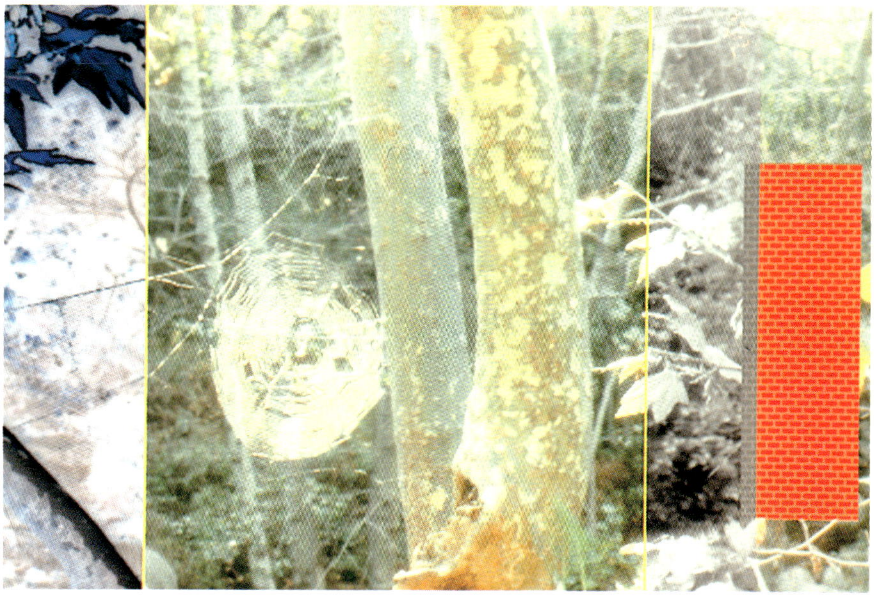

COLOR PLATE 1 Some elements used to make digital photograph of Color Plate 2. Image made with TIPS software and QFX utility program.

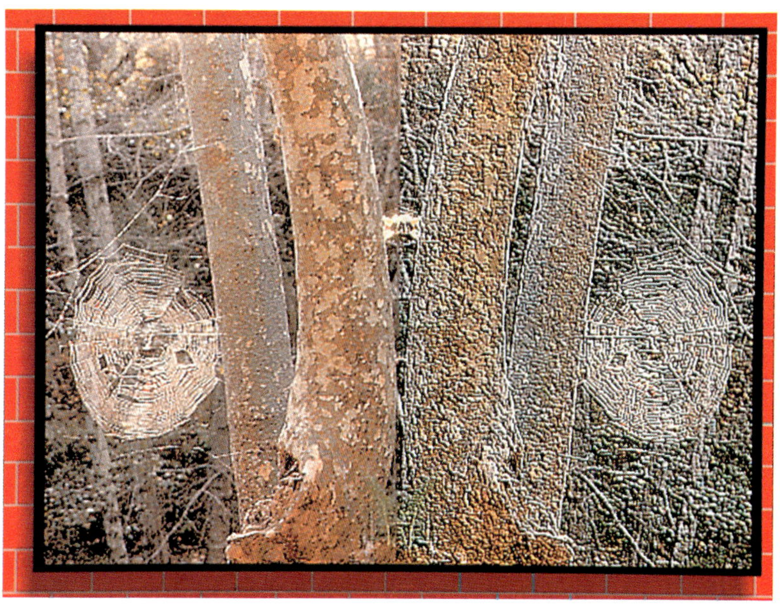

COLOR PLATE 2 Trees in Color Plate 1, framed in black, and hung on a wall. Made with TIPS and LUMENA and utility software, from an Ektachrome 200 slide.

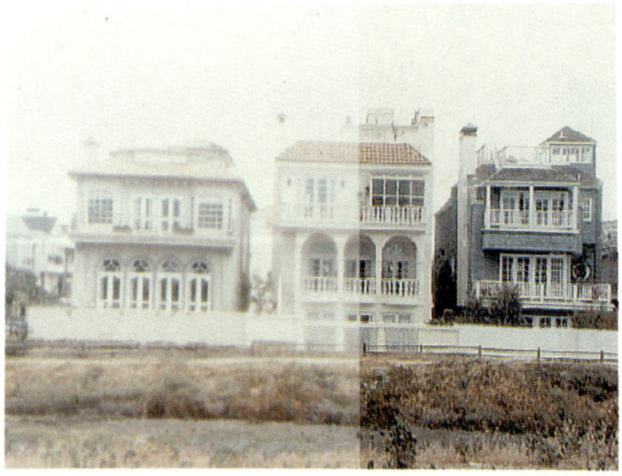

COLOR PLATE 3 Original image used to make digital photographs of Color Plates 4 and 5. Shot on Ektachrome 200 slide film. The left side of the image is overexposed about 1.5 stops, the right side was brightness and color corrected using TIPS and QFX utility program.

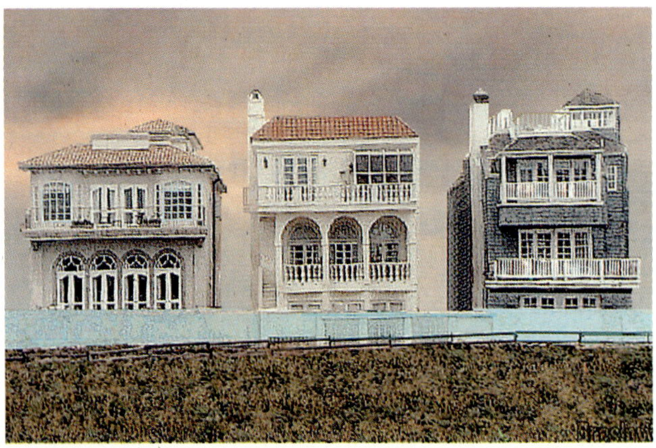

COLOR PLATE 4 Made from digital photograph of Color Plate 3 with TIPS and utility programs.

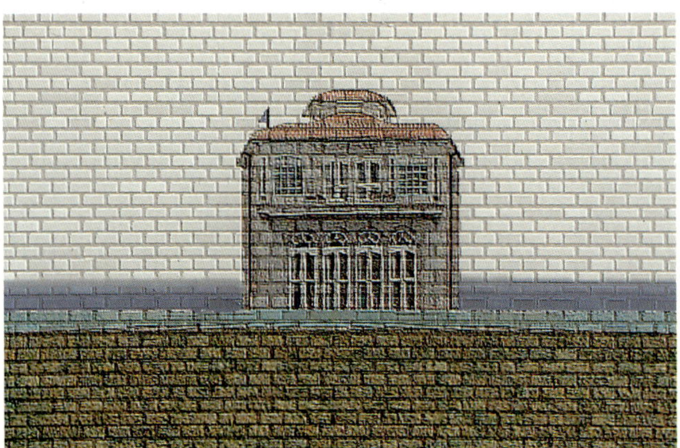

COLOR PLATE 5 Made from digital photograph of Color Plate 4, using TIPS software.

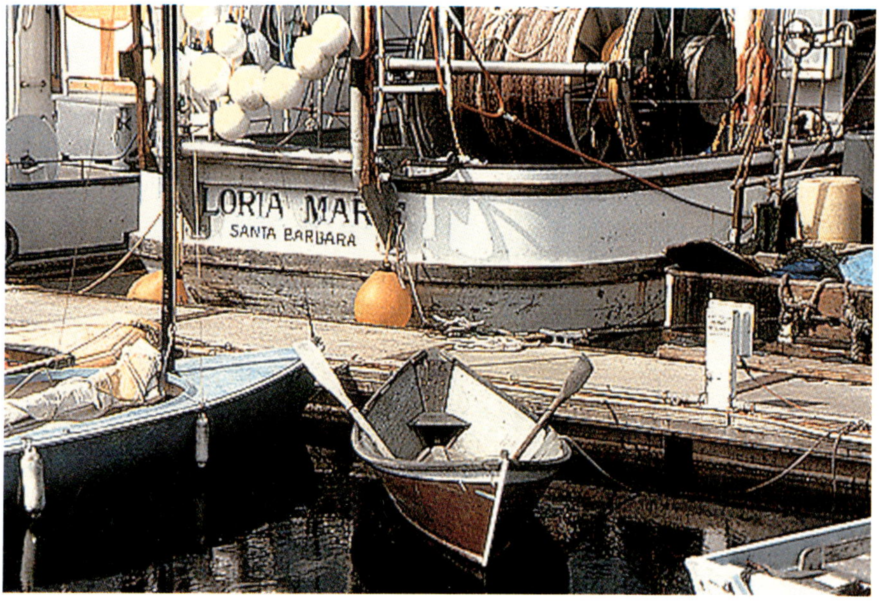

COLOR PLATE 6 The general quality of a digital photograph, shot on Ektachrome 200 slide film, and made with LUMENA.

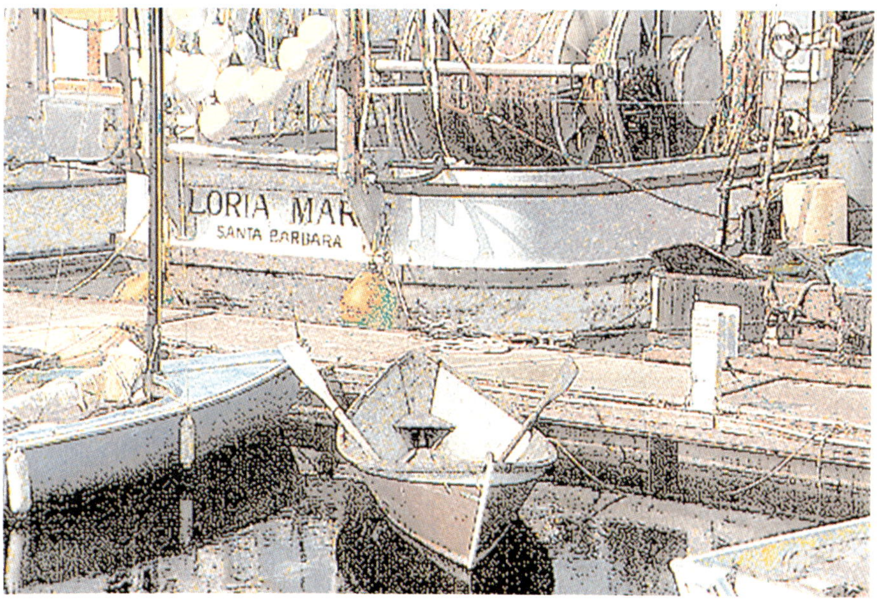

COLOR PLATE 7 A pastel image made with LUMENA software from digital photograph of Color Plate 6.

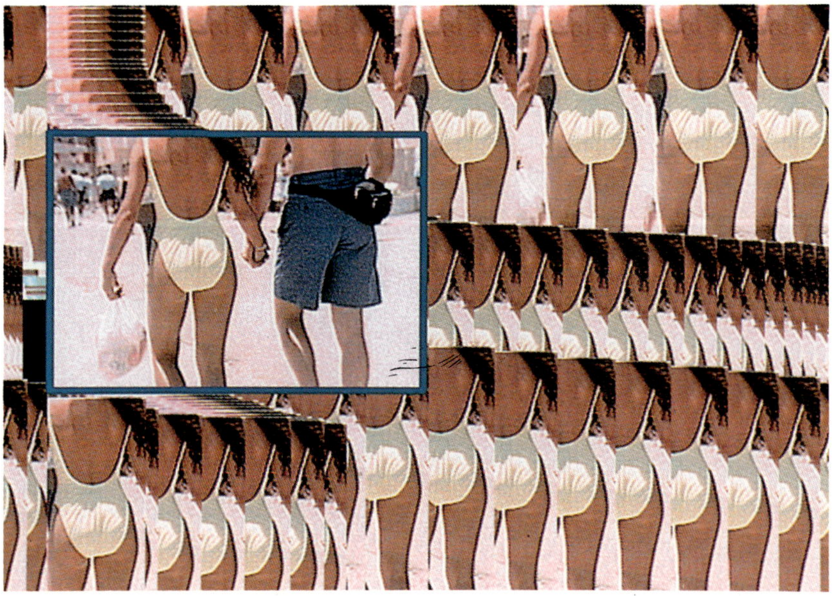

COLOR PLATE 8 Made using a Tamron Fotovix II-X slide transfer device from an Ektachrome 200 slide, with the image manipulated by TIPS software.

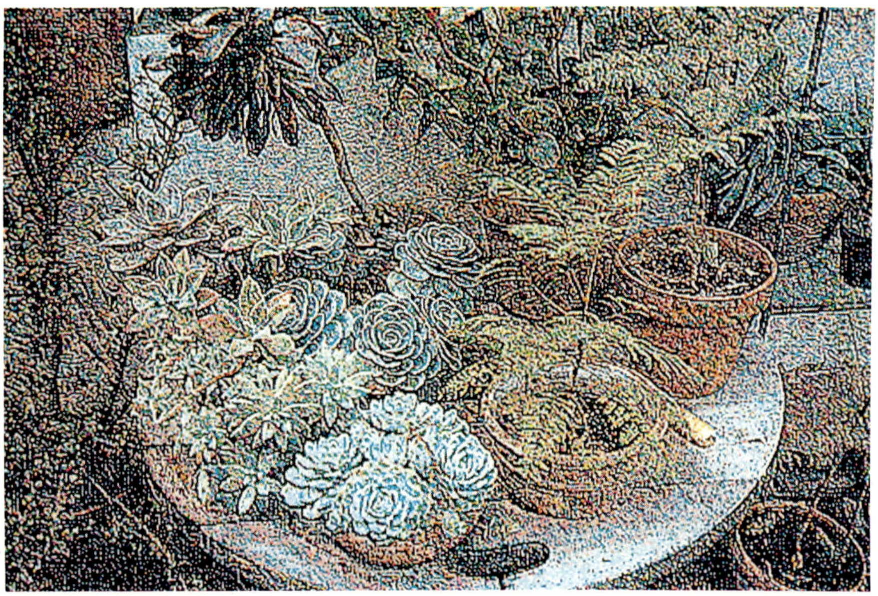

COLOR PLATE 9 Made with LUMENA software, from an Ektachrome 200 slide.

After you have manipulated the image with your paint program, you have the option of outputting it to either a film recorder or paper printer. At present, film recorders will produce the highest quality, especially at 4K or higher resolutions. Keep in mind that 35mm slides will yield better results at 2K resolution than 4 x 5 inch transparency film at 4K resolution.

By the early 1990s, paper printers, especially those working on dye sublimation methods, such as that used by the Sony XU-5000, and thermal dye transfer such as that used by the Kodak XL-7700, will probably be the output method of choice for many digital photographers. While dye sublimation and similar processes currently yield true photographic quality as far as tonal ranges are concerned, the image contains many jaggies. This is because the printer does not take the image from a high-resolution file (e.g., a large scanned file or a large pixel interpolated file), but rather from the image that appears on the graphics monitor. The image generated on the monitor contains too few pixels to eliminate jaggies. Kodak's thermal dye transfer printers accept input from files, but the image is limited to 2048 x 1536 (in 35mm ratio) pixels, spread out over an 8.5 inch by 11 inch area, which may not be enough to completely eliminate jaggies. However, printers of this type will be extremely popular. With them you will be able to get a continuous scale color or B&W print in a matter of a few minutes, without any darkroom work needed, and the quality will increase as technology improves.

The information given in this book will enable you to communicate with sellers of digital photography equipment. Before you buy any equipment discussed in this book, you should discuss the *entire* system with a graphics professional. What scanners are available for the board you are interested in? What paint programs? What utility programs? What film recorders? What choices are there in drivers? Will an 800 TV line gray-scale video camera yield good results? If you get the answers to these questions before you buy any equipment, you should be able to avoid any disappointments due to incompatibility of equipment.

DIGITAL AND TRADITIONAL PHOTOGRAPHY COSTS COMPARED

Depending on the computer, graphics board, monitor, and program(s) you buy, and whether you also purchase other equipment such as a scanner or film recorder, a functional digital photography system could cost from $3,000 to $25,000 or more. It may seem that digital photography equipment is more expensive than equipment used in traditional photography, but the following comparison of prices will show that this is not the case. For this comparison, I will categorize the digital photography equipment focused on in this book as being at the middle-upper price range of what is available. The digital photography equipment I have been recommending represents the middle-to-upper range of prices. (I have not recommended the more expensive Vista 4M or NuVista 4M systems because the increase in quality of these systems is not commensurate with their increase in cost.) Moreover, the prices I have given are, for the most part, list prices, and it may be possible to purchase some of this equipment discounted. For our purposes, I have selected current traditional photography prices from mail order ads (discounted prices) for some middle-upper price range traditional photographic equipment for comparative purposes. While it is true that lower cost traditional photography equipment is available, the ones I have selected are by no means the most expensive, just as the digital photography equipment I have recommended is by no means the most expensive.

1. *35mm format equipment*: Nikon F4 camera body, $2,000; 24mm f/2.0 lens, $500; 50–300mm zoom f/4.5 ED lens, $2,150; 500mm f/4.0 ED-IF lens, $3,500. Total cost of this camera and three-lens system: $8,150.
2. *2 1/4 inch format equipment*: Hasselblad 553 ELX camera body, $1,600; film back, $420; 40mm f/4.0 CF lens, $2,450; 80mm f/2.8 CF lens, $925; 140–280mm CF zoom lens, $4,000. Total cost of this three-lens camera system: $9,395.
3. *4 x 5 inch format equipment*: Sinar P2 4 x 5, $5,000; digital shutter, $3,350; 90mm f/5.6 Schneider Super Angulon lens; $1,000; 210mm f/5.6 Schneider Symmar-S lens, $550; 360mm f/6.5 Nikkor-W lens, $1400. Total cost of this three-lens camera system: $11,300.

With each of these systems, I have excluded additional equipment which most photographers would have, such as wide-angle bellows for the Sinar view camera, or extra film backs for the Hasselblad, which could raise the overall prices considerably.

This comparison illustrates two points. First, the cost of the digital photography equipment is within the same range as that of comparable traditional photography equipment. Second, there is very little overlap in the required equipment for these two methods of photography. For example, in digital photography you don't need a view camera, since you can correct perspective with the LUMENA, ColorScheme II, or TIPS perspective functions. Expensive lenses are not necessary in digital photography, since the finished image is a result of manipulating numbers, which means that the image can be sharpened and pseudodetail added (both impossible to do with traditional photography).

Often traditional photographers have dealt with the problem of the high cost of photographic equipment by pooling resources and "going in" on expensive equipment together. This works well in theory: Each co-owner now has access to a necessary piece of equipment. However, in practice, problems arise, such as when two (or more) co-owners need the same $2,000 lens for a day's shooting. And unless co-owners of studio strobes and other lighting equipment share the same studio space, moving heavy/bulky/delicate lighting equipment around from one studio to another can get tiresome.

Co-owning a scanner is a different story. Since it only takes about five minutes to scan a slide or print, two, three, or ten photographers can co-own one scanner and rarely get in each other's way. The biggest problem may be in agreeing on a central location to house the scanner, although they are easily transportable. Film recorders can also be co-owned, but many will have to be left at a central location, since they can be quite heavy. I do not suggest that computers be co-owned because the time you spend in front of your monitor trying out different ideas can cause even the best of friends to stop speaking to each other. As digital photography becomes more popular, it will be easier to find other photographers with whom to split the cost of equipment, or those who do own equipment can act as service bureaus for those who don't, and thus recoup part of the cost of the equipment.

A final point which must be made is that the price of computer equipment tends to rapidly decrease, while the quality of the equipment rapidly increases. In 1986 I bought a monochrome IBM PC with a 4.77MHz clock speed, two 360K floppy disk drives, and a 20MB hard disk for about $2,750. In 1988 I bought a monochrome CompuAdd 286/8, with clock speeds of 8MHz and 6MHz, one 1.2MB, one 360K floppy disk drive, and a 40MB hard drive, for about $1,750. In 1989 my wife bought a monochrome CompuAdd 286/12, with clock speeds of 12MHz and 6MHz, one 1.2MB floppy drive, and a 40MB hard drive, for about $1,600. With each successive system, the cost went down and the abilities of the computer increased. It is safe to assume that this will be the case with the equipment mentioned in this book.

PART III

FINISHING TOUCHES

CHAPTER 10

HOW THE PICTURES WERE MADE

So far we have been discussing digital photography in very abstract terms. In this chapter we will put these generalizations into practice. First the specific equipment used to create the images appearing in this book will be described, and then I will explain how each image was made. It is not my intention to try to teach you to duplicate these images, but rather to give you a sense of what is involved in making digital photographs of varying complexity. Finally, there will be a brief summary.

EQUIPMENT USED

The computer equipment used to make the digital images that appear in this book consisted of a CompuAdd 286/8 AT-compatible computer operating at a clock speed of 8MHz, a monochrome Hercules compatible, standard graphics expansion board, a monochrome monitor, a TARGA-16 digital photography graphics board, and an Electrohome 1311 long persistence, phosphor color monitor. The computer also had one 1.2MB floppy disk drive, one 360K floppy disk drive, one 40MB Seagate 251-1 hard drive, an Intel AboveBoard meeting the LIM standard with 2MB of RAM configured as expanded memory, and a Microsoft serial Mouse.

The programs used to make the images included TIPS-16, LUMENA-16, QFX, and Norman Breslow's Photo Utilities.

All the digital photographs which began as traditional photographs were shot with a Nikon N-2000 with a 35–135mm Nikkor f/3.5 zoom lens, except for one which used a Nikkor 50mm f/2 lens, with either Tri-X (ISO 400) black and white film or Ektachrome (ISO 200) color slide film.

The input devices used to put the images into the computer varied from image to image, and will be described individually. All the image files were sent to

Digital Image Company, a service bureau, where they were used to generate slides using a Matrix QCR-Z film recorder at 2K resolution.

HOW THE IMAGES WERE MADE

Both TIPS and LUMENA were discussed extensively in Chapter 7. Some information pertinent to this chapter follows: When using TIPS, a small icon menu is used to select effects and their parameters. This is done by calling the icon menu to the screen, which obscures part of any image that is already on the monitor screen. A color to be used by the program can be selected by either placing the cursor on the color in the color palette in the icon menu, or by placing the cursor on a color in the image surrounding the icon menu. A drawing function, such as a rectangle, can be selected by placing the cursor on the icon representing a rectangle. To create the desired effect, the icon menu is removed from the monitor screen. At this point, the image on the monitor screen can be manipulated.

LUMENA places a menu on the standard computer monitor, and not on the digital photography graphics monitor as TIPS does. Tools and parameters are selected by moving the cursor to an extreme edge of the graphics monitor screen: The cursor disappears from the graphics monitor and appears on the standard monitor. After making selections from the menu, the cursor is moved to the extreme edge of the standard monitor screen: It disappears and reappears on the graphics monitor screen, where the cursor is used to manipulate the image.

Figure 2.1 was originally shot with Tri-X film, and shows the computer used to make the images in this book. A black and white print was scanned to 512 x 342 resolution by Jetgraphix.

The original image contained a drafting lamp positioned between the Electrohome 1311 monitor on the left and the computer and standard monitor to the right. The lamp was removed by using TIPS drawing functions. First, the cursor size and shape were chosen. The most frequently used shapes are probably "dots" or filled circles. The smallest dot-shaped cursor is the size of one pixel, and the largest is the size of about 15 pixels in diameter. A small dot is good for working with very small elements, such as an eyelash, but larger cursors are better when working with large areas: Since they cover more area, they make changes more rapidly. A medium size dot cursor was selected. The *freehand drawing* function was chosen from the icon menu. When in this mode, anything the cursor touches is turned into the color the computer was set to use. Then, the color of the wall in the digitized image was selected, and this color became the color the computer was set to use. Once the cursor size, color, and the freehand drawing function were chosen, the cancel button on the mouse was pressed to remove the icon menu from the monitor screen. The cursor was positioned over the lamp, and by moving the cursor over the lamp in the digitized image, while pressing down on the select button of the mouse, the lamp was removed and replaced with the color of the wall.

Unfortunately, the bottom of the lamp was attached to the table the computer was on. Using the wall color to remove that part of the lamp would change the lighter table to the color of the wall. The icon menu was retrieved (by pushing the cancel button on the mouse), the cursor was positioned on the table in the digitized image, and the select button on the mouse was pressed. The working color changed from that of the wall to that of the table. The icon menu was removed from the monitor screen by pressing the cancel button, and the cursor was positioned over the part of the lamp that was attached to the table. Pressing down on the select button, the cursor was moved over the lamp, thus replacing the lamp with the color of the table.

When the lamp was completely removed from the image, the icon menu was recalled, and the *move* function was chosen. As mentioned in Chapter 7, the move function in TIPS can do a number of things, including moving a portion of the image. A rectangle was placed around the graphics monitor (the monitor on the left in the image), and the monitor was repositioned to make it look like it was floating in the air. This created two minor problems. First, part of the table the monitor was sitting on was now also floating in the air, where it shouldn't have been, and part of the table became white, lighter than the color of the table, because the area where the monitor had been was replaced with white. These problems were easily solved. The offending white area on the table was changed to the color of the table (a light gray) by calling up the icon menu and again selecting the freehand drawing function, then removing the icon menu from the monitor screen, and finally moving the cursor over the white area, which was replaced with the color of the table. To remove the small section of the table that was floating in the air just under the graphics monitor, I selected the wall color again, and moved the cursor under the monitor while pressing the select button on the mouse.

At this point I had removed a lamp from the image, and floated a monitor in the air. I then decided to see what the image would look like if it were slightly posterized. TIPS does not have a posterization effect, although other programs, such as QFX and LUMENA, do. I chose to use one of my own posterization programs, which yields similar effects to those provided by either QFX or LUMENA. To do this I exited (left) TIPS. (Although I was no longer in TIPS, the image I had been working on remained on my graphics monitor. The graphics monitor works independently of TIPS, or put another way, TIPS is used to make changes to the RAM used by the graphics board. The graphics board will accept instructions from any program written to work with it, and will continue to display the last image on the monitor even after the program has stopped working. Therefore, even though I left TIPS, the image I was working on was still on the graphics monitor.) I selected a black and white posterization program from my utility program, the standard graphics monitor displayed a menu asking me to enter the degree of posterization I wanted to use, I selected a light posterization, and the computer changed the image on the graphics monitor to a posterized one, in about five seconds.

Once this was done, I typed the command to reenter TIPS. The posterized image was on the graphics monitor, and I could now use the effects in TIPS to continue to manipulate the image. I decided that the color of the wall was too close to that of the table. I selected a dark gray from the color palette on the icon menu, and then selected the *fill* function, which can be used to rapidly change an area made up of one color to another color. I put the cursor on the color of the wall in the image, and pressed the select button on the mouse, and the lighter wall color was changed to the working color, which was the dark gray I had just selected. I decided to lighten the top of the desk, and did so by using the fill function again, after selecting a very light gray from the icon menu.

The final step was to create the letters and lines that appear on the image. TIPS has about 14 different fonts (complete alphabets) to choose from. I selected a font and color (black). When you are in lettering mode, TIPS asks you to draw a line on the monitor where the type should be placed. You do this by moving the cursor to the beginning and end points of the line you want to create, and pressing select at each extreme. The cursor then moves to the beginning of the line, and you enter the lettering with the computer keyboard. As you type, the letters in the font you selected appear on the line. If you want, you can press <cr> and the cursor will drop beneath the original line, and allow you to type another line.

To create the letters in the image, I created a short line where the letter "A" was to go, I typed an uppercase A, pressed <cr> and typed B, then repeated this for the letters C and D. I repositioned the line for the letter E and again for the letter F. I then called up the icon menu to select the *line* function, and a small cursor. The line function allows you to draw a straight line by indicating the starting and ending points of the line. To do this, I placed the cursor to the right of the letter A, pressed and released select, and then moved the cursor to the right. As I moved the cursor, a line followed. When I was satisfied with the line, I pressed select again, and the line was made permanent. The thickness of the line depends on the size of the cursor chosen. I repeated this procedure to construct lines as needed.

The image was completed and needed to be saved and stored in a file. I used the appropriate function from the icon menu, typed a file name, and pressed <cr>. The image was sent to the hard disk as a file, taking up about 410KB. When the time came to have the digital image made into a slide, I copied the file from the hard disk to a 1.2MB floppy, and mailed it off, with a completed order form, to Digital Image. A few days later I received the slide, along with the floppy disk.

Figure 2.2 was made by shooting a computer chip, which was placed on a photographic gray card, with +3 and +2 diopters on the front of a 50mm f/2 Nikkor lens. The computer chip was very small, measuring about 1/2 inch in length, by about 1/4 inch in depth. The resulting image on the Tri-X negative covered only about 1/16 of the entire film area. The negative was placed in a Fotovix II-X slide transfer device, which was attached to the TARGA-16 graphics board by a standard video cable. The Fotovix II-X was set to convert a negative image into a positive image, and the zoom lens was set to its maximum 6X magnification. TIPS was used as the paint program and was set, using the icon menu, to its video capture mode. With the image correctly positioned on the screen, the select button on the mouse was pressed, and 1/30 of a second later the image was digitized.

This image was not sharp, which was not the fault of the Fotovix II-X, but resulted from greatly enlarging an image with little depth of field. To try to correct this problem I exited TIPS and entered Ron Scotts QFX utility program, which has a sharpening function. I selected that feature, and QFX asked me to select the amount of sharpening to be done: light, medium, or heavy. I selected light and then moved the mouse (which also works with QFX) to the word "execute" on the menu used by QFX. When the sharpening function began, I was asked to draw a rectangle around the area of the image to be sharpened. I drew a rectangle around the left half of the image, and then pressed select. The program sharpened only the left side of the image. I then retrieved the main menu, by pressing cancel, and reset the sharpening function to work at medium. This time I sharpened the right side of the image.

When you look at the left side of the image, you will notice that it is not sharp, and that the gray card the chip was placed on seems mottled. The right side is sharper, but mottled horizontal lines are evident. These lines are the *raster scan lines* that make up a video image. If a video image is sharpened too much, these lines become evident. On the left, lightly sharpened side, the lines are just beginning to become visible, which causes the mottled effect. This example illustrates that although images can be sharpened, this function is not free of problems. If the image is very blurry to begin with, sharpening may not work, and it definitely won't work with a very unsharp video image.

Figure 2.3 started off as a Tri-X negative, from which a print was made, and then scanned at a resolution of 512 x 342 by Jetgraphix. Some electrical wires were visible along the right side of the image, and they were removed with TIPS, using the same procedures described in Figure 2.1. The upper part of the computer case was also changed to the gray that appears in the photograph using both the filled rectangle function and freehand drawing function. Finally, the entire image was sharpened using the sharpening function in QFX, set to the lightest setting. Even this setting proved to be too sharp: Some of the highlight area (reflections of light) on the floppy disk "broke up," giving a spotted effect, and part of the hand also became a bit mottled. Both of these problems could have been easily corrected by going back into TIPS and using the freehand drawing tools to "retouch" these areas, but I left them untouched to show you some of the potential consequences of sharpening. The lettering was added using the TIPS lettering function, as described in Figure 2.1.

Figure 2.4 was also shot on Tri-X film. The print was scanned at a 512 x 342 resolution by Jetgraphix. The color of the area at the top of the image was changed to a darker gray using the filled rectangle function in TIPS. The image was sharpened with QFX, and then TIPS was used again to create the letters.

Figure 4.1 was shot with a Sony ProMavica still video camera. The original color image was converted to black and white using my utility programs. The major problem with this image was that due to the low contrast scale of the video camera, exasperated by the digitizing function of the TARGA board, the dark cars in the foreground lacked detail, as did the floor of the car showroom. The column in the center of the image looks realistic, since it was neither too light nor too dark for the ProMavica to handle. Using the freehand drawing function helped define the car on the left, by smoothing out some highlight reflections. The image was sharpened with QFX, at its lightest setting, which again caused some mottling on the black car and showroom floor. Part of this mottling results from the original image being video, but sometimes the lightest setting on QFX's sharpen function is too strong. LUMENA's sharpen function has a much broader range of sharpening, which eliminates the problem of oversharpening, although raster scan lines of video images still present a problem.

Notice the jaggies throughout the entire image. The jaggies are particularly noticeable because this image was not scanned, but *grabbed* by the TARGA board. More jaggies will appear from a grabbed video image than from a scanned image. I could have been "kinder" and selected an image that would have made the ProMavica look better, but I wanted you to see what I consider to be a typical result from a normal sunlit scene. However, if you choose the subject with care and light it carefully with flat light, better quality images can be obtained when using the ProMavica.

The two rectangles that indicate the image area which will appear on one complete monitor screen at 1K resolution (large rectangle) and 2K resolution (small rectangle), discussed in Chapter 7, were added to the finished image with TIPS. White was chosen from the color palette, and the unfilled rectangle was chosen from the icon menu, so that when the rectangle was completed, only its outline became white, and everything inside the rectangle remained unchanged.

Figure 4.2 was made using the drawing functions in TIPS. If you look at the circle at the bottom of the image, and the triangle at the top, you will notice that both have dark grays on one side that gradually change to light grays. These effects are created with TIPS's *spread* function. You choose two colors, or two shades of gray, and tell TIPS to create a transition from the first color to the second. You can then use this range of colors in the same way you would use one color. For example, the circle on the bottom of the drawing was created by choosing *circle* from the icon drawing menu, and then selecting the *filled* option. With the spread function also chosen, a circle appeared filled with the range of colors in the spread palette. The triangle at the top of the image was created using the same procedure with the triangle function. In both cases the resulting effect is that of a three-dimensional object, sphere or cone, rather than a two-dimensional object, circle or triangle. The black lines were placed in the image with the *line* function, as described for Figure 2.1, and the different shades of gray were created using the filled rectangle function with different grays chosen from the color palette.

Figure 7.1 began as a video image from a Canon E-708 consumer grade video camera, grabbed using TIPS. The original image was similar to the lower left portion of the photograph, although it filled the entire screen of the graphics monitor. Using the *move* function in TIPS, the original image was reduced to the size of one of the small rectangles in the image. Using the *tile* function, the small image of the mouse was duplicated throughout the screen, filling it with 25 small mice. To break up the pattern, I enlarged one of the 25 mice, using the move function from the icon menu, and repositioned the result at the lower left side of the image. Another mouse was selected, and slightly enlarged, and then a third mouse was selected, but this time reversed (flopped) left to right. This last procedure involved once again selecting the move function, but this time using its reverse ability, instead of its enlarge/reduce ability. This procedure involved drawing a rectangle around the area to be reversed, and then pressing an icon that represents flopping the image left to right, and the computer took care of the rest.

Figure 8.1 was shot with Tri-X film, printed, and scanned at 512 x 342 resolution by Jetgraphix. The resulting image was heavily posterized using my utilities, and then saved to a file. During the posterization process the lettering "Bright," "Contrast," and "Power" began to merge with the color of the monitor's cabinet. The lettering that appears on the monitor's case is the result of my trying to trace the parts of the letters that remained after the posterization. Although computers can do a lot, you will find that some drawing skills are sometimes necessary, and if you have limited drawing skills, the computer won't be able to cover for you. The problem of the missing letters could have be handled in at least three different ways: The lettering could have been done with TIPS' or LUMENA's fonts; the original image could have been reposterized at a lower setting, which would have protected the lettering; or a window file of the letters could have been made prior to posterizing the image, then pasted onto the posterized image. The freehand drawing function could then be used to blend the non-posterized colors surrounding the lettering in the .win file with the posterized monitor case.

Figure 8.2 was photographed off the graphics monitor screen as described in Chapter 8. The original image of the cat was shot on 8mm videotape, and then digitized using TIPS video capture function. Colors at various parts of the cat were blended with adjoining colors, using the TIPS *blend* function. Blend allows you to blend one color with another, which creates watercolor or blurred effects, depending on the degree of blend selected. Finally, the "bars" were added to the image from the TIPS *pattern* menu. This menu includes 16 predrawn patterns. All the patterns have transparent backgrounds, so that they can be overlaid onto an existing image, allowing the existing image to show through.

Figure 10.1 was shot on Ektachrome film, and the slide was scanned by Barneyscan. The 512 x 342 image was then changed to black and white using my utilities, although either QFX or LUMENA could have been used instead.

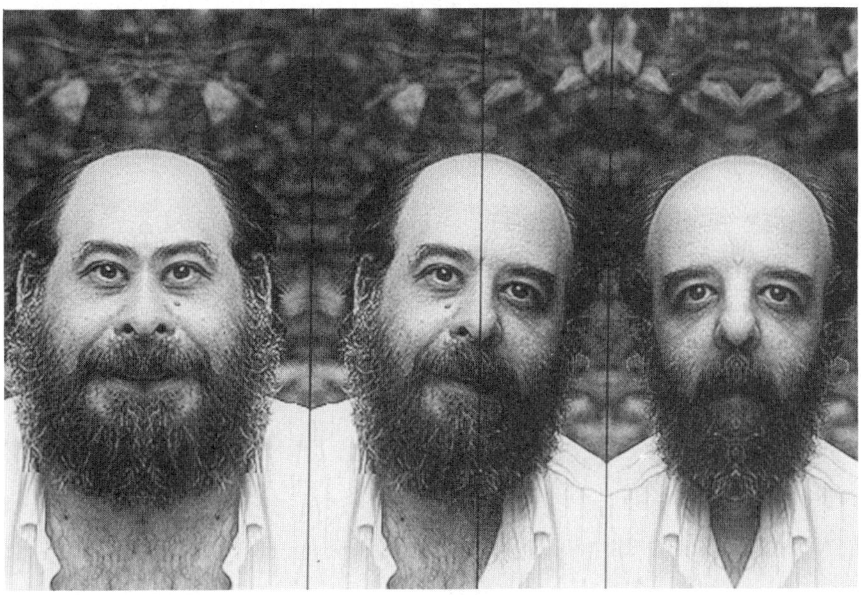

FIGURE 10.1 A man's face, photographed with a Nikon-2000 with Ektachrome 200 film, which has been manipulated using TIPS software and utility programs.

You will notice that there is a large black rectangle surrounding the head in the center of the image, with a black vertical line running through it, dividing it in two. The image within the large black rectangle was on the original slide, with some additional picture area on either side of it. Figure 10.1 was made in the following way: The TIPS move function was chosen, and a rectangle, matching the smaller left rectangle, was drawn. This isolated the right side of the face (the viewer's left). That part of the image was moved to the extreme left side of the monitor screen, leaving the original face in place. I then pressed select, which *pasted* the right side of the face permanently into position, and then moved the partial face toward the center of the image. I now had a complete face in the center of the picture, the right side of the face at the extreme left side of the monitor, and the same partial face between the two. Using the flop capabilities of the move function, I reversed the partial face, and then attached it to the partial face on the left of the monitor. This created a new complete face, made up of two right sides of the original face. The same procedure was carried out for the left side of the man's face, (the viewer's right), creating a new face made up of two left sides of the original face. A mole on the left side of the face was removed using the free-hand drawing tool, to make the two sides of that face look slightly different. The reason the three faces all look different is because faces are not symmetrical, with the left and right sides usually being quite different. If the head is tilted, the asymmetry is exaggerated. Finally, I exited TIPS and used a utility program to change the color image into a gray scale image.

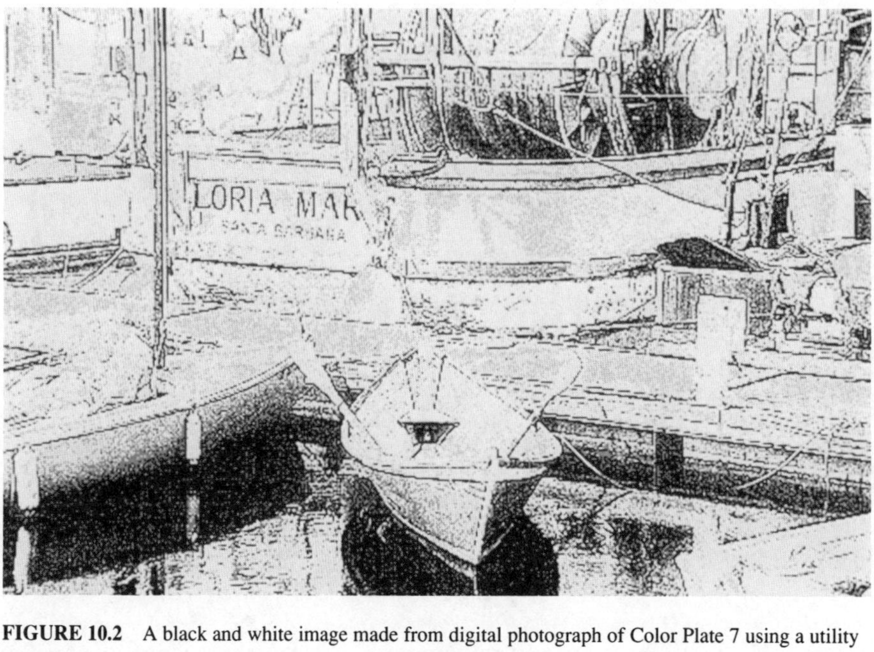

FIGURE 10.2 A black and white image made from digital photograph of Color Plate 7 using a utility program.

Figure 10.2 is a black and white image made from the digital photograph of Color Plate 7, which will be explained later in this chapter.

Color Plate 1 shows a wooded scene with a large spider web. The original 35mm Ektachrome slide was scanned by Jetgraphix to 512 x 342 resolution. The large central area shows the wooded scene, which has a slightly green cast and is soft in focus. The extreme left side of the image shows part of the picture area, made into a negative. This was done to illustrate that the colors of all or part of an image can be reversed, and to describe using a window file.

To make the color reversal, the entire image was first displayed on the graphics monitor using TIPS. TIPS does not have a function to reverse the colors in an image, but LUMENA, QFX, and other utility programs do. After the image was displayed, I exited TIPS and used my utilities to reverse the colors in the image. I then reentered TIPS and gave the instruction to make a window file. TIPS asked me to select the area of the image on the monitor I wanted to make into the .win file. I drew a rectangle around the area on the left of the image, then gave the window file a name, and TIPS created the file. I then told TIPS to put the original (not color-reversed) image on the monitor. (Remember, when you work on an image, you are making changes in RAM, and the original image file is not altered. You can alter the original image file by *saving* a new or changed image to a file with the same name as the original image file. This is called *overwriting* an image file.) TIPS put the original wood scene on the graphics monitor, covering up the image with reversed colors. I then instructed TIPS to display the window file. A few seconds later the reversed color image I had stored as a .win file was placed over the left side of the original picture on the monitor. I then saved this entire image as a paint file, called "woods1".

On the right side of Color Plate 1 are some red bricks (a part of which has been changed to gray). The image of the bricks is included in TIPS (and LUMENA) as *texture files*. These textures can be incorporated into photographs or drawings. I displayed the bricks on the graphics monitor, by pressing the bricks icon in the icon menu, and selecting the filled rectangle icon option. I then drew a rectangle on the right side of the picture, and the rectangle filled with the red bricks. To illustrate that you can continually make changes to digital images, I decided to turn part of the color image into a gray scale. To do this I left TIPS, changed the entire image to a gray scale as briefly described in Figure 10.1, then used TIPS to make a window file of a small portion of the now gray scale woods and bricks. I then merged the image file "woods1" with the window file. Finally, I selected a yellow color from the color palette, and, with the line function from the icon menu, created the two yellow lines dividing the different areas of the image.

Color Plate 2 was generated using some of the elements discussed in the previous example. First, the original file containing the wooded scene was displayed on the graphics monitor, using the *read* image file function in QFX. Then, using QFX, the image was color corrected to eliminate the green cast. The light sharpening command of QFX resulted in a sharp image without the problems encountered when sharpening video images described above. After this, I exited QFX, used LUMENA to take part of the left side of the image, and moved it to the right side of the monitor screen, thus covering up part of the wooded scene, and then flopped it over, creating a mirror image. I embossed the right side of the image, having set LUMENA's embossing tool level at about half strength. Then I drew a black rectangle around the scene, so as to get the effect of a picture frame. At this point I left LUMENA and called up TIPS.

Once in TIPS, I stored the color-corrected, sharpened, mirror-imaged, and embossed picture as a window file. I then used the filled rectangle and the bricks texture file and created a rectangle about half the size of the monitor screen. I then enlarged this image using the move/enlarge function until the monitor was filled with the now larger bricks. I then called up the window file of the wood scene, which appeared on the center of the monitor, surrounded by the bricks.

I decided to create a slight drop shadow to give some depth to the entire image. To do this, I used the shadow effect in QFX. (The shadow that appears in this image could have been darker or lighter, wider or narrower, and could have been positioned on the right side of the framed picture, or on the top or bottom.)

Color Plate 3 shows a not particularly interesting image made on Ektachrome slide film. It was about 1.5 stops overexposed, and scanned by Jetgraphix to a 512 x 342 resolution file. The left side of the photograph shows the overexposed image, and the right side shows it corrected for brightness, contrast, and color using effects in QFX.

Color Plate 4 was made from the slide in Color Plate 3. This image took a full day to complete, and a full description would be rather lengthy. The seven major manipulations involved: (1) Color and contrast corrections of the original image were made with QFX. (2) Removal of the sky and shrubbery, utility poles, and other distractions surrounding the houses, which was accomplished using the freehand drawing function in TIPS. The shrubbery that was along the sides of some of the houses was removed by "coloring over" the shrubbery with the color and texture of the houses. The sky was covered with black (opaqued), using the filled rectangle and freehand drawing function. (3) The grass was made thicker by using the *dupe* feature in TIPS, discussed in Chapter 7. One area of grass was duplicated on another area of the image, then another area of grass was selected and duplicated somewhere else, etc., until the grass was thicker. (4) The small wooden fence was enhanced using the freehand drawing functions, as were the color and texture of the larger concrete wall. (5) The houses were repositioned using the move function of TIPS. (6) The entire image was then modified, using my utilities, to make it appear to be a painting. (7) A new sky was added to the image by merging the image of the houses, as a window file, to the image of a sunset, which had been shot on 8mm videotape and then digitized.

You may notice that part of the sky looks slightly mottled, while the non-sky area doesn't. This occurred when I sharpened the image. The sharpening function works well on the scanned portion of the image, but the sky was originally a video image, and as stated earlier, these do not sharpen very successfully. When the image was complete, it was made into a slide by Digital Image. I requested that the 512 x 342 file be pixel *replicated* (not interpolated) to 2K resolution. If you look at the curved window areas of the houses, you will notice they have a rather "blocky" appearance, caused by the replication. A smoother image would have resulted if interpolation were used.

Color Plate 5 was made using the house on the left in Color Plate 4. (1) The desired house was saved into a window file, the screen cleared of all information (it went white), and then the .win file was redisplayed on the monitor screen. (2) The right edge of the house was flopped and copied to the left edge of the house; the shape of the roof was changed by removing some parts and moving others. (3) The house was made to look taller by adding a flopped section of the lower story of the house to an area beneath the house. This covered the wall, giving the illusion that the house was taller than it was. (4) The grass was raised up toward the house and leveled. (5) A videotaped sky at sunset replaced the original sky using the opaquing and merging functions. (6) The "ocean" was drawn using the airbrush functions and line drawing functions in TIPS. The image of the seaside house at sunset was saved as a paint file. (7) A second image file was made of a wall made up of white bricks. The bricks from Color Plate 2 were colored white and then made to look like a weathered wall by introducing some irregularities into the shapes of the bricks, and then using the shadow effect in QFX to make them look more realistic.

I wanted to create the effect of a scene painted on a brick wall. To do this I entered LUMENA. I saved the house file into LUMENA's buffer (which is a portion of RAM in which images can be temporarily stored) and then had LUMENA display the bricks on the monitor screen. Using LUMENA's transparency function set at 50% I merged the image of the house from the buffer with the image of the brick wall that was on the monitor. At a setting of 50%, the effect was similar to that of placing one slide over another one. This "sandwiched slide" effect caused the image of the house stored in the buffer to blend together with the image of the bricks that was on the monitor screen. I then saved this final image to a file.

The photograph on the cover was originally shot with a consumer grade video camera, digitized with a Targa-16 board, and transformed in about 90 seconds with Norman Breslow's Painting Effects ($400).

Color Plate 6 was shot on Ektachrome film and scanned by Barneyscan to a 512 x 342 resolution file. The image was not enhanced, and is shown to give you an idea of what an average quality, medium-resolution scanned image looks like. Depending on the quality of the original slide or print, your medium-resolution scanned images may look similar to this, or better or worse. As we saw in Color Plate 3, computer enhancement can successfully modify slides or prints that are about 1.5 stops under- or overexposed, so this image could easily be modified (made lighter or darker, or more or less contrasty, etc.).

Color Plate 7 was made from Color Plate 6, using LUMENA. First, the image was stored as a buffer. Then LUMENA's colorize tool was selected. This tool was mentioned briefly in Chapter 7, and results in an image that looks as if it were made using pastel chalk. I told LUMENA to display the image that was in the buffer, and the colorize tool changed it into a pastel-looking image. Abandoning the colorize tool, I selected the transparency mode, setting the level of transparency to 30%. Finally, I displayed the original image once again, which remained stored in the buffer, combining it with the pastel image on the screen. Because the transparency setting was only 30%, the pastels were more visible than the rich colors of the original image. If the setting had been 70%, the richer colors of the image stored in the buffer would have been dominant.

Color Plate 8 was shot on Ektachrome film and then transferred to the TARGA-16 board using a Fotovix II-X slide transfer device. The image contained in the original slide is displayed in the box. This image was reduced in size using the TIPS move function, and then stored in a window file in TIPS.

Although the image had been stored as a window file, it still remained on the graphics monitor. I used the move function again, choosing its *paint* option, and then drew a rectangle around the woman. When move/paint is selected, the selected image can be moved around the monitor screen, and *pasted* on the screen as it is moved. The effect shown in Color Plate 8 is only one of the many possible effects this function can make. When I had finished with the paint option, I called the window file of the original image to the screen and pasted it in position. Using the rectangle function, I created a blue frame around the window image to set it off from the rest of the image. Finally, the image was sharpened with QFX.

Color Plate 9 was shot on Ektachrome film, and scanned by Barneyscan to 512 x 342 resolution. LUMENA was used to create a *color-embossed* effect. This effect is different from the *embossed* effect used for Color Plate 2, and gives unpredictable results when used with full scale color photographic images. The new, color-embossed image was combined with the original image using the transparency tool, as described in Color Plate 7.

SUMMARY

In this chapter I have tried to give you a feeling for how images can be manipulated using a computer. You will not be limited to using only one program, just as you are not limited to using one lens or one type of film when shooting traditional photographs. If you need to select one program, LUMENA does have the most options. In fact, most of the photographs in this book could have been made solely with LUMENA. If you use TIPS, utility programs will be necessary, since TIPS does not presently contain posterization, sharpening, or extensive color correction abilities. But even if you have LUMENA, you may want to use other programs because you may find that a particular effect is achieved in a manner more to your liking by these other programs.

I used my utilities extensively for the images in this book, but probably all of the images could have been made with LUMENA, or LUMENA and QFX. Some of the images, such as those involving *transparency* and *colorize*, could not have been made without LUMENA. TIPS could not have been used to make some of these images, even with QFX or other utilities.

Another point should be apparent from the description of the processes by which these images have been created. Many of the effects illustrated resulted from much experimentation and a certain amount of serendipity. Working with high resolution files would make this virtually impossible. Not only does working with large files pose logistical problems (e.g., imagine going through the steps to create Color Plate 4 on four separate sections of a 1K resolution file, or on 16 different sections of a 2K resolution file), but whatever effect you manage on one portion you will need to be able to duplicate on the others. This essentially precludes "playing around" with an image because each step will need to be documented to ensure your ability to obtain *exactly* the same look in each section.

CHAPTER 11

THE FUTURE OF DIGITAL PHOTOGRAPHY

In this chapter I will discuss some esthetic considerations relating to digital photography, the effects of high technology on digital photography, and some legal and ethical issues. Finally, some suggestions for continuing your digital photography education will be offered.

THE ESTHETICS OF DIGITAL PHOTOGRAPHY

Every photographer who has taken formal courses in photography, or who has read books on the subject, knows that a series of rules have been developed to help define what is, and what is not, a "good" photograph. For instance, a black and white photographic print should contain a photographic black (the darkest black the photographic paper is capable of depicting) and a photographic white (the color of the photographic paper being used). Additionally, the print should contain shadow detail, highlight detail, and a continuous range of grays between the photographic black and white. Prints that meet these requirements are said to have proper contrast, and look "natural," while prints that do not are said to be either "flat" (low contrast) or "too contrasty."

In the preface to this book I stated that there were no rules in digital photography, simply because the imagery is too new for the genre to have become regularized/standardized. Throughout this book, however, I have made reference to quality in digital photography, and have stated that some products were capable of creating quality images, while others were not. If there are no rules in digital photography, what does quality in digital photography mean, and how can it be judged? If there really are no rules, how can some products be said to create quality images, while others can't?

It is inevitable that a digital photograph will be compared, at least initially, to a standard "traditional" photograph. For this reason I have discussed quality in terms of whether particular equipment is capable of creating an image that compares favorably with a quality traditional photograph (as defined by the rules of traditional photography). Time will be needed to find out whether digital photography should be judged by the rules of traditional photography or not. If digital photography is not to be judged according to the same rules as traditional photography, then will new rules be generated? Or will the broad range of possibilities afforded by digital photography preclude one system of rules?

One area I have emphasized in discussing quality in digital photography concerns *jaggies*. Jaggies are the only telltale evidence that a photographic image was created with a computer. I have given the impression that jaggies are undesirable. If you don't want your images to show signs of computer manipulations, then they are. But jaggies may well become accepted as a natural by-product of digital photography, (at least during the current state of the art of digital photography), and some people may even try to incorporate or emphasize jaggies in their overall design. It is too soon to know whether the "rules of digital photography" will state that the image should be jaggie free. Similarly, it is too soon to tell whether a *thermal transfer print* made up of noticeable dots will be considered acceptable, or whether the digital photographer will be taught to avoid them at all costs. But if we apply the rules of traditional photography to digital photography, then both jaggies and prints made up of noticeable dots will be considered inferior, since both yield images that do not show continuous tonal changes throughout the image. If you are capable of suspending your knowledge of the rules of traditional photography, you will probably be less inhibited in your approach to digital photography.

HIGH TECHNOLOGY AND DIGITAL PHOTOGRAPHY

In the preface of this book I also mentioned that imagery in traditional photography was a product of technology: Advances in lens design changed the look of traditional photography, as did faster (more light sensitive) film. The state of technology at any given time has a direct bearing on what traditional photographs look like. Although some changes have been introduced on the basis of insight of some photographers, most major innovations resulted from technologic advances: The traditional photographer waits for the development of new technology, and then uses the new technology to create new imagery; the photographer is dependent on technologic advances to push photography into new visual territory.

Throughout this book I have stated that digital photography is a product of technology. Nonetheless I have also stated that the range of possible applications and types of images are limited only by the imagination of the photographer. Why is it that the traditional photographer is constrained within the boundaries defined by the available technology, while the digital photographer faces few such barriers even though, if anything, the technology is even more present in the process?

There are at least two reasons. The first has to do with the range of choice in types of equipment with which the photographer may choose to work. The technology of traditional photography is in the hands of relatively few companies. It takes tens, if not hundreds of millions of dollars to create a new type of color film and the chemicals necessary to process it, or to design new lenses or other equipment. In digital photography, numerous companies are involved in research and development. The development of new digital photography hardware products often entails combining readily available state-of-the-art computer chips in novel ways. Because no single company, or small group of companies, is in control of the digital photography technology or determines what is or is not possible, the choices open to the digital photographer are greater than the choices available to the traditional photographer.

A second reason for the increased freedom of digital photography has to do with the user's own involvement in the manipulation of the process. Most digital photography hardware is *programmable*. This means that you can write your own programs to manipulate the equipment in unique ways. *Many* books are sold that teach computer programming (although I suggest that you take a few formal courses). You don't have to be a computer genius to learn to program.

You can get an idea of the control *you* have over the visuals you create, by considering my utility programs. My utility program includes an "inferred" effect. I have not seen this effect in any other program, so the effect is unique. If I did not make the program available to others, I would be the only person capable of creating this effect (for the time being). You could write an "inferred" program too. However, it is very unlikely that the results of your "inferred" effect would look exactly like mine. It may look better! And this "inferred" effect is only one of countless effects that are possible. Thus, because the technology of digital photography is different from the technology of traditional photography, you don't have to wait for a company to come out with new products to help you create unique visuals; you can come out with your own products.

SOME ISSUES OF LAW AND ETHICS

Just as digital photographic technology promises a tremendous diversity of new imagery, it also creates the opportunity for the misuse of copyright laws. A copyright is the legal right to reproduce and sell materials ranging from photographs to poetry. The material contained in this book, including the photographs, are copyrighted.

The problem arises because scanner technology will allow anyone with the proper equipment to digitize the photographs, manipulate them, and offer the altered images for sale as their own.

Ordinarily, copying the photographs, without expressed permission from the copyright holder, is illegal. If you made a photocopy of the images, for your own use, you would be breaking the copyright law, but it is doubtful that you would be prosecuted. If you made photocopies and offered them for sale, and if the copyright holder found out about it, you probably would be prosecuted.

But altering an image raises questions not as yet adjudicated. The question may well be, how different is your image from the original? At what point has something completely new been created? An attorney would argue that the original image was copyrighted, and hence anything derived from it is a violation of the copyright. The success of this argument may well depend on the extent to which the new image differs from the original, and we will have to wait for a number of cases involving digitized copyright infringements to have a better idea of what will be permissible.

It is inevitable that such lawsuits will occur, because the temptation is just too great. For example, an advertising agency could save thousands, if not tens of thousands of dollars, on location shots. Why pay to have a model, photographer, and an entourage fly to Hawaii, Tahiti, or Jamaica when there is an almost unlimited supply of scenics of these locations? A scenic on a picture postcard could be scanned, the sky could be changed, some rocks moved to different locations, and a model or product, shot in a studio, could be merged with the altered image. If the owner of the copyright to the original postcard recognized the altered image as her original image, she could sue for copyright infringement, but the outcome of the lawsuit is far from certain.

Other types of situations will also arise. What if a photographer shoots a job for one client, and then discovers that he could use part of the image for another client? Or what about the photographer who scans other photographer's work from magazines, and uses these images as part of his own portfolio? The potential for misuse of the images made by others is great. While this is a legal issue that will be resolved, it is also a moral issue. One thing to keep in mind is that no one is immune: You can appropriate someone else's copyrighted material, but the same can be done to yours just as easily. Perhaps the best guideline, until legal guidelines emerge, is "don't do unto others as you would not have them do unto you."

CONTINUING YOUR DIGITAL PHOTOGRAPHY EDUCATION

The competition in the computer marketplace is great, and new products are constantly being marketed. You must keep informed about the changes that occur. The best way to do this is to subscribe to computer graphics magazines, a number of which are listed in the Sources section. As digital photography becomes more popular, more magazines will appear. For now, most of these magazines use a shotgun approach to their articles. They cover animation, methods used to merge architectural models of a shopping center with a photograph of the actual location, means of combining type with an image, etc. None exists solely for those interested in still digital photography. However, the articles will teach you about the equipment available and how it is used, and you will be able to determine its value in relation to your interests and needs. What may be of most value are the listings of new products which can be found in these magazines.

There is one publication that concentrates on still digital photography, *Digital Photography Newsletter*. It is not a magazine, but a non-technical newsletter written for beginning and intermediate digital photographers. It describes new products—how they are used, and what their value is to the digital photographer—and also discusses unique ways to use existing digital photography equipment. While I can highly recommend *Digital Photography Newsletter*, I suggest that you also subscribe to a more traditional computer magazine, for a broader awareness of trends and insights.

UPDATES

In the six months between the time this manuscript was completed and the actual publishing process began, a number of changes have occurred.

1. CompuAdd is a large mail order company that sells its own line of computers, as well as peripherals made by other manufacturers, and the manufacturer of the computer used to make the photographs in this book. They are listed under *mail order* in the "Sources" section. CompuAdd has redesigned its lower end (less expensive) computers. They now come with a host of features built in, such as serial and parallel ports, LIMS compatibility, and the ability to add up to 4MB of extended or expanded RAM onto the motherboard. However, CompuAdd has reduced the number of expansion slots from eight to five in these less expensive computers. Furthermore, two of the five slots will only accept "half-size" expansion boards, which are shorter than the standard "full-size" boards. Apparently, CompuAdd's reasoning is that you will not need as many expansion slots since many features that ordinarily used these slots are built into the motherboard. However, in digital photography, many non-standard expansion boards, such as GPIBs, are needed. While CompuAdd is a good source for computer supplies at reduced mail order prices, you should seriously consider whether a computer that has only three full-size expansion slots is a good deal in the long run, regardless of its current low selling price. Because there are so many variables involved in the purchase of a computer to be used in digital photography, I strongly suggest that you purchase one from an authorized dealer in digital photography equipment (see the Sources section).

2. Lasergraphics has announced a $1000 price reduction in its LFR models.

3. Truevision announced a reduction in the surcharge for their Vista and NuVista series boards, due to a decrease in the cost of RAM chips. The 1M now has a suggested list price of $2,995, the 2M lists for $4,495, and the 4M for $6,495.

4. Tamron has come out with the Fotovix III ($999), which is a stripped down version of the II-X. This new model is limited to 35mm format, has a 3X zoom instead of a 6X zoom, and does not have the complete range of color correction capabilities that the II-X does. At its low price, it may warrant consideration for an input device.

5. A version of Ron Scott's QFX special effects program is now available for Vista Boards.

6. Sharp has announced price reductions on its scanners. The JX-450 now lists for $6,995, and the JX-300 has a list price of $3,995. Additionally, Sharp now sells GPIBs for $570, and offers free scanning software (driver) with the purchase of one of its scanners.

7. Howtek now bundles a GPIB and software with its Scanmaster ($8,200) and Scanmaster II ($6,200) color scanners.

8. Kodak has announced a list price of $18,000 for its B&W, and $25,000 for its color XL-7700 Thermal Dry Transfer Printer. The cost per color print is said to be about $3 each. Although at the upper end of equipment mentioned in this book, these printers should compete well with such film recorders as the Matrix QCR-Z and the Solitaire-8 and Solitaire-16. For those who want a dry, 8.5 x 11 inch print in less than six minutes, and who would ordinarily consider a higher quality film recorder, this printer should be seriously considered.

9. Matrox, a manufacturer of quality expansion graphics boards used mostly for CADD, has introduced the Illuminator-16 ($1,500), which is a 100% TARGA-16 clone. According to Matrox, any program that runs on the TARGA-16 graphics board will run on the Illuminator-16. This differs from the Vision-16 board, which does not work with software written for TARGA boards but needs software specially written for it. The Illuminator-16 comes in versions for IBM/AT and IBM-compatibles, and IBM PS/2 computers, in versions for NTSC or PAL video standards. An expanded model (about $2,500) works with 1K resolution TARGA images, but this version is presently limited in software capable of running TARGA files at this increased resolution. Also of note is the Illuminator-16's ability to add special effects to incoming or outgoing video signals (e.g., make posterizations, mosaics), but this feature will not work on images when they are being painted. For your nearest authorized dealer, contact Matrox at 1055 S. Regis Blvd., Dorval, Que, H9P 2T4 Canada, (514) 685-2630, and in the United States, phone 1 (800) 361-4903.

10. Mathematica has announced a paint program called TEMPRA ($595), designed to work with the Illuminator-16 graphics board. TEMPRA is a very close cousin to TIPS, and has a number of improvements that go beyond TIPS' capabilities. TEMPRA can work on images of up to 8K resolution, in sections similar to LUMENA's HI RES, will work at 1K resolution with the version of Illuminator-16 which can display 1K resolution files, and will read and write .tga (TARGA) files as well as .tif (used widely in desktop publishing, and a standard file type created by almost all scanners) and .pcx (a standard EGA/VGA file type). TEMPRA is extremely versatile. It is available both from Matrox, or from Mathematica at 402 South Kentucky Ave., Suite 210, Lakeland, Florida 33801. Tel. 1 (800) 852-MATH.

11. Sable is the same program as TEMPRA, but is has been rewritten to work only with the TARGA-16 board, and not with the the Illuminator-16 board. TEMPRA has been made to only work with the Illuminator-16 board, and not with the TARGA-16 board. Sable is sold through AT&T Graphics Software Labs.

12. Hercules, a very well known and respected manufacturer of computer graphics boards, has introduced the Hercules Graphics Station Card ($1,024), which is really a number of graphics boards in one. It functions as an EGA/VGA graphics board, a Super VGA board, and a Super TARGA-16-type board (32K simultaneously displayed colors at 640 x 480 pixel resolution) and a TARGA-24-type board (16.7M simultaneously displayed colors at 640 x 480 resolution). At present, there is no software for the Super TARGA-16 mode, and only a modified version of LUMENA-32 for its TARGA-24-like mode. A special version of LUMENA and the Hercules Graphics Station Board are sold together for $3,995, or about the same as a TARGA-24 board without any software, or they can be purchased separately. The only deficiency in the Hercules Graphics Station Card is that it does not include a frame grabber, so those who want to shoot/grab video will have to purchase a separate stand-alone video grabber board (from a few hundred to a few thousand dollars). For more information, contact Hercules at Hercules Computer Technology, 921 Parker Street, Berkeley, California 94710. Tel. 1 (800) 532-0600.

The introduction of lower priced graphics boards which have greater capabilities than the standard TARGA-16 and 24 boards is the crack in the dam of dominance by Truevision's TARGA and Vista boards. They must now either lower the prices of these boards and/or improve them to compete with not only the Illuminator-16 and Hercules Graphics Station Card, but also with other boards that are sure to follow. Similarly, programs such as TIPS will have to have lower prices and/or improved capabilities, to compete with programs such as the lower priced TEMPRA/Sable and other programs that are sure to be introduced in the future. By the time you begin to put your digital photography system together, expect to pay considerably less, and get considerably more, than the prices and features of equipment reported in this book. Some of those who sell Truevision (TARGA) products do not feel threatened by the introduction of lower priced new equipment for a number of reasons. First, they too can sell these competing boards and programs, and second, they feel that the low list prices (and probable lower street prices) mean that the profit margins will be so low that little, if any, customer support will be available. The reasoning is that you pay more for Truevision products, but you get support (help in the beginning getting the products to operate correctly, and answers to questions as they arise thereafter), which you won't get when you buy these lower priced competing products. The question is, if you don't need support, do you need to spend more money and get fewer features? If you have been using a TARGA-16 board, and you want to get a TARGA-24 or 32, do you feel confident enough to buy a Hercules

Graphics Station Card, use their customer support line if you run into problems getting it up and running, and then rely on your own experience to learn how to properly use the equipment, and save a few thousand dollars in the process? For many, the answer will be a resounding "Yes."

The price of color scanners is also coming down, which will be appreciated by most entering into digital photography. Earlier I mentioned that joint ownership of scanning equipment or sending slides/prints off to service bureaus were two methods of beating the high cost of scanning. Over the past six months a development not contemplated has arisen. Desktop publishing centers have been opening in meaningful numbers in larger cities. These are often offshoots of photocopy/instant-print businesses, which install a number of computers with word processing programs suitable for designing ads, catalogs, or newletters. Many of these desktop publishing businesses also have B&W and color scanners, with low charges (often from $5 to $15) per scan. As soon as these businesses add film recorders and/or sophisticated printers such as the Kodak XL-7700 to their list of services, the ubiquitous digital photo lab will be closer to reality. As things currently stand, it will not be the traditional photolabs that make the transition to digital photography, but the desktop publishing business centers that bring this about.

It should be obvious from these updates that digital photography is a very fast moving industry, and you should take advantage of publications such as *Digital Photography Newsletter* and some of the magazines found in the Sources section to keep abreast of changes as they occur.

ADDENDUM

I have just reviewed the typeset version of this book, and thus have the opportunity to add this important news. Truevision has announced that it is discontinuing it's TARGA series of boards and replacing them with one new board called the TARGA Plus. However, all the information concerning TARGA and the use of TARGA boards and compatible products is still valid, since the new board is 100% compatible with all existing TARGA software.

The TARGA Plus incorporates the TARGA-8, M8, and 16 models for $1,795, and for $1,995 will include the TARGA-32. It will display all these versions at standard resolution (512 x 482 or fewer pixels) and will also be capable of displaying them at 1k (1024 x 764) resolution. This means that it will be possible to manipulate 1K scanned files as one screen, instead of the previous need to break the image into four screens. Additionally, the board is said to have greatly improved video capabilities, although it is doubtful whether the images will be as good as 1K scanned files. However, the improved video capabilities are said to be valuable to those interested in animation, as well as television stations.

The TARGA Plus comes in versions for the IBM AT and compatibles, or for the PS/2s, and versions with either NTSC, RGB and S-Video, or PAL, RGB and S-Video.

This shows that prices do come down, and quality goes up. For about the price of a TARGA-16 outfit outlined in Chapter 9, you can have a TARGA-32, which operates at 1K resolution.

GLOSSARY

8088 The designation of the Intel computer chip which the earliest IBM/PC/XTs and PC/XT-compatibles were designed around.

80286 or **286** The designation of the Intel computer chip which the IBM/AT and AT-compatibles were designed around.

80386 or **386** The computer chip "386" computers are designed around.

<cr> Abbreviation for carriage return found in manuals and books on computers. It means to press the carriage return key (sometimes labeled "Enter") on the computer keyboard.

.BAT A file extension, short for Batch. A *Batch file* is a list of instructions the computer will carry out in order. The computer acts as if you were typing information into it. A Batch program can be used to run a film recorder at night, by giving it the names of the files you want printed. When the film recorder finishes with one file, it looks to the Batch file for the name of the next file to use.

.COM A file extension the computer recognizes as identifying a program file.

.EXE A file extension the computer recognizes as identifying a program file.

Access Time The average time it takes the read/write head to move from track to track on a hard disk looking for information. These times usually vary from one type of hard disk to another. The fastest which are usually found on AT-type computers have an access time of about 16ms (milliseconds), with the slowest around 65ms.

Addressable Resolution The maximum number of pixels the graphics expansion board is capable of manipulating. This may or may not be the same amount the graphics board or monitor is capable of displaying. See *displayable resolution*.

Adaptor Board Same as *expansion board*.

Architecture (1) The physical design of a computer. (2) The philosophical design of a computer. (3) Both the physical and philosophical design of a computer.

Background Mode In some cases, IBM/ATs and AT-compatibles are capable of performing certain functions while also being used for another purpose. For example, you may be able to work on a digital photograph while making a print of another one at the same time "in the background."

Batch Files See *.BAT*.

Baud A unit of measurement of the speed of a modem transmission. The higher the baud, the faster the transmission. Most modems now operate at 1200 or 2400 baud, although 9600 and higher baud modems are currently available. As a rule of thumb, you can think of baud as bits per second transmission speed. Divide the baud by 8 and you will have the number of bytes transferred per second.

Bays Physical space inside the computer for floppy and hard drives.

BIOS Abbreviation for Basic Input/Output System, the part of the computer that oversees communications between the computer and peripherals such as the keyboard, monitor, and I/O ports.

Bit One eighth of a byte. A bit is either a positive or negative electric charge. The sequence of eight bits (on/off states) creates one of 256 different bytes ($2^8 = 256$).

Bundling (1) The combination of two or more different functions onto one expansion board or peripheral device. (2) The practice of including something free with a product as an inducement to buy the product. For example, an EGA paint program that ordinarily costs $100 may be included free when you by a particular mouse.

Bus A network of wires which expansion boards and specialized computer chips use to communicate with each other. At present, for digital photography, the bus speed, or speed that the communications take place, must be 8MHz or slower.

Byte Eight bits. Depending on the organization of the eight bits, one byte is interpreted by the computer as a character, digit, punctuation mark, or specialized keyboard symbol. One or more bytes are interpreted by a computer graphics board as a color.

CCD Abbreviation for Charged Coupled Device. A computer chip with light receptors (e.g., pixels) used in many video cameras and scanners.

CGA Abbreviation for Color Graphics Adaptor, the first IBM/PC color graphics standard.

Chip An integrated circuit consisting of etched pathways that carry electric charges. The design of the pathways determines what the chip is capable of doing. Some computer chips are designed for communications purposes, others for computer memory, etc.

Clock Speed The speed at which the CPU is capable of looking for instructions to execute. Most IBM/AT-compatible computers have a clock speed of 10, 12, 16, or 20MHz (megahertz), and a safe clock speed of either 6 or 8MHz.

Clone Computer equipment that works like other, usually well established, computer equipment. A clone can be either inferior or superior to the original equipment it is modeled after.

Closed Architecture A computer that cannot be easily modified or upgraded after it leaves the factory.

Color Graphics Controller An expansion board that puts an image on a color monitor.

Compatibility The ability of two or more computer products to work with one another.

Configured Hardware or software that is put into proper working order. Hardware and software may have to be configured differently for different computers. Instructions will be found with the product.

Control Key A special key on computer keyboards. Its use is outlined in computer program manuals.

Copy Protected Copies cannot be (easily) made from copy protected software. This protects software manufacturers from unauthorized copies of their programs being used.

CPU The central processing unit of a computer. The computer's "brain."

Cursor A bar of light on the computer monitor that shows where the next letter from a word processing program, or the next action of a paint program, will take place.

Default When computer programs or peripherals are first turned on, they usually have certain functions automatically set. These default settings can usually be changed by the user as the need arises.

Desktop Publishing Using computers and peripherals to publish magazines, newsletters, sales information, etc.

Digital Photography Changing a photograph into numbers so the image can be manipulated with a computer, and then changing the numbers back into a photographic image.

Digitizing Tablet A device for moving the cursor around a monitor screen. Good for both digital photography and drawing, but considerably more expensive than a mouse.

Disk See *floppy disk, hard disk.*

Diskette Same as *floppy disk.*

Disk Drive (1) Same as *floppy disk drive.* (2) Same as *hard disk drive.*

Disk Drive Controller The computer mechanism that positions the read/write head(s) to take information from or put information on a floppy or hard disk.

Disk Operating System A series of programs that are used to control basic computer functions.

Displayable Resolution The number of pixels the graphics expansion board is capable of lighting on a graphics monitor. See *addressable resolution.*

Dithered A strategy to create the look of more colors or shades of gray when using a printer which is limited in the number of colors or shades of gray it is capable of making, by arranging light and dark colors/grays in such a manner as to cause an optical illusion. Dithered prints do not look photographic.

DOS Abbreviation for Disk Operating System, it is a series of programs needed to run an IBM/PC/XT/AT computer. See *MSDOS, PCDOS.*

Double Sided (1) A floppy drive that reads and writes to both sides of a floppy disk. (2) A floppy disk that is capable of having information written to or read from both of its sides.

Downwardly Compatible When new high-tech computer equipment is designed to also work with older, lower tech equipment, it is downwardly compatible.

Drive Specifier The letter (A: B: C: D:, etc.) associated with a disk drive.

Driver A special program that works as an intermediary between hardware and software. Some drivers help run a mouse or tablet, film recorders, scanners, etc.

EGA Abbreviation for Enhanced Graphics Adaptor. A mid-resolution, very limited number of simultaneously displayed colors, IBM and IBM-compatible graphics display system, good for charts and graphs.

EMS Abbreviation for Expanded or Extended Memory Standard. See *LIMS.*

Emulation The ability of software or hardware to mimic some other software or hardware.

Expanded Memory A method for fooling the computer into using more than the maximum 640K of RAM most personal computers are designed to use.

Expansion Board A circuit board that fits into an expansion slot located on the motherboard. Some adaptor boards come installed in the computer, and others may be purchased at a later time to upgrade or add abilities to the computer.

Expansion Slots Female slots located on the motherboard into which the male pins of an expansion board fit.

Extended Memory A method of using RAM as a disk. See *RAM disk.*

External Modem A modem that resides outside the computer, and that attaches to the computer's serial port by a cable.

Field One half of the scan lines (either the odd or even ones) that make up a complete video image. One field is one half of a frame.

Fixed Disk Same as *hard disk.*

Floppy Same as *floppy disk.*

Floppy Disk A physical device that can be used to permanently store computer information on magnetic media. Floppy disks can be removed from the computer for storage or transportation purposes.

Floppy Disk Drive The mechanism that contains the read/write heads which take information from or put information on a floppy disk.

Floppy Diskette Same as *floppy disk.*

Frame A complete video image made up of two fields.

Frame Grabber An expansion board with the ability to digitize a videotape or TV signal for use in digital photography.

Frequency The signal, in megahertz, at which computer equipment, such as the computer's clock, works.

Fully Populated Contains the maximum amount of (usually) RAM the expansion board or peripheral device is capable of holding.

Full Height The original size of a floppy drive. See *half height.*

Function Keys Keys labeled F1 through F10 (or higher) on a computer keyboard. Program designers sometimes enable these keys to perform certain tasks. For example, pressing F1 when using certain programs will cause a help menu to be displayed on the monitor.

GB Abbreviation for gigabyte.

Gigabyte 1,073,741,824 bytes (often rounded to one billion bytes).

GPIB General Purpose Interchange Bus. A special type of I/O board used by some scanners, film recorders, and printers.

Graphics Expansion Board An expansion board that puts graphics (pictures/charts) on a monitor.

Half Height A floppy or hard drive one half the size of a full-height drive. The half-height drive is now the standard size for floppy drives, and for most hard drives with capacities of 65MB or less.

Hardware Physical parts of computers, such as expansion boards, or attachments such as printers, monitors, etc. See *software.*

Hard Disk A device used to store large amounts of information. Unlike a floppy disk, the magnetic media is on a metal base, which is not removable from the hard disk drive. The hard disk(s) and hard drive (e.g., the read/write heads) are one unit, and the two terms are often used interchangeably.

Hard Drive See *hard disk.*

Help Menu With many programs, a list of suggestions can be displayed on a monitor at the user's option. Often, help menus are context sensitive. This means that the program senses what you are attempting to do, and an appropriate help menu will appear if you ask for it.

Hue Color.

ICB Acronym for Image Capture Board, a low-resolution, moderately high simultaneously displayable color, graphics expansion board. Not well suited for digital photography.

Intensity Lightness or darkness of a color.

Internal Modem A modem that resides inside the computer, and contains its own serial port.

I/O Board Abbreviation for Input/Output board, which usually has serial and parallel connectors that cables attach to. Information from the computer can be sent out to other devices through this board, and information from devices can be sent into the computer through it.

Jaggies Diagonal and circular lines that appear jagged or like a staircase, because there are not enough pixels to make them look smooth.

K The binary number 1024.

KB Abbreviation for kilobyte.

Kilobyte 1024 bytes (often rounded to 1000 bytes).

LIMS Acronym for Lotus Intel Microsoft Standard for expanded memory and extended memory.

Logged Drive The drive the computer is associated with at any particular time.

Long Persistence Phosphors Refers to monitors that attempt to reduce annoying flickering of the image by having the luminescent phosphors which make up the image glow for a long time. These monitors are used for digital photography purposes, are not generally available through mail order, and tend to be costly. See *short persistence phosphors.*

LUMENA A paint program used in digital photography.

M In this book, M stands for million.

Mass Storage Devices Disk drives capable of storing large amounts of information, such as a hard disk.

MB Abbreviation for megabyte.

Megabyte 1,048,576 bytes (often rounded to 1,000,000 bytes).

MDA Abbreviation for monochrome display adaptor.

Menu A list of alternatives.

Modem A method used to turn computer information into frequencies that can be transmitted over telephone lines, where a modem on the other end changes the frequencies back into computer information. From the words *MOdulate DEModulate.*

Monitor The television-like device on which information from the computer appears.

Monochrome With computers, black and one color, usually amber, green, and less often, white.

Monochrome Display Adaptor or **Board** (1) Originally, an expansion board that displayed text on a computer monochrome monitor. (2) Today, most MDAs will also put graphics/drawings on a monochrome monitor .

Motherboard That part of the computer that contains the expansion slots into which expansion boards are placed, and usually the CPU, and other computer chips necessary to make the computer work.

Mouse A device for moving the cursor around the monitor. Very good for digital photography, not so good for drawing.

MSDOS Abbreviation for Microsoft Disk Operating System. A group of programs needed to run IBM/PC/XT/AT-compatible computers. PCDOS and MSDOS are functionally the same, although PCDOS can only be used with IBMs and MSDOS only with compatibles. See *PCDOS, DOS.*

Multifunction Board A type of expansion board that performs a number of diverse tasks or functions, such as both controlling a monitor and serving as an I/O board.

Multisync A device, usually a monitor, that is capable of receiving signals over a broad range of frequencies.

NTSC Acronym for National Television Standards Committee. The standard that United States television works under. See *PAL.*

OCR Abbreviation for Optical Character Recognition. Many scanners can be used to read typeset letters and convert them for use in word processing or other types of programs.

Open Architecture A computer capable of accepting equipment made by a variety of manufacturers to expand the computer's original capabilities.

Optical Drives Removable mass storage device that is similar to a WORM, since both use a laser beam to write information to the disk, but differs in that information can be erased from optical drives.

Overlay The ability of most TARGA and compatible boards to merge a live video image with a painted or digitized photographic image.

Paint Program The type of program used to manipulate digitized photographs, or to paint or draw images.

PAL Abbreviation for Phase Alternation Line, the standard for European television signals.

Palette (1) The total number of colors a graphics expansion board is capable of manipulating. See *total palette.* (2) The total number of colors the graphics expansion board is capable of displaying on the monitor at one time. See *simultaneously displayed palette.*

Panning Some graphics expansion boards can manipulate more information than it, or the computer monitor, is capable of displaying (see *addressable resolution* and *displayable resolution*). If this is the case, it is possible to put information held in the graphics board onto the screen by removing other information that is already on the screen. This is done by moving the cursor to the left of the screen which will display any information that exists to the left but that is not displayed, and removes information on the right of the screen to make room for it. Any information to the right of the screen can similarly be displayed.

Parallel Transmission A method by which the computer sends or receives eight bits simultaneously. As soon as the eight bits are received, the computer interprets them as a byte. See *serial transmission.*

PC Abbreviation for Personal Computer, which is a computer used by one person at a time. Other classes of computers, such as mainframes or minicomputers, can be used by a number of people at one time.

PCDOS The disk operating system for the IBM PC, XT, and AT computers, but not compatibles. See *DOS, MSDOS.*

Peripherals Computerese for auxiliary attachments.

Personal Computer See *PC.*

Physical Disk A disk or drive that physically exists, as opposed to a RAM disk that exists only as electrical charges in RAM chips.

Pixel Short for picture element (PIX EL), the smallest amount of information that can appear on a monitor.

Pixel Interpolation A method for increasing the size of a file by creating pixels. With this method, three pixels created between a black and a white will be dark gray, middle tone gray, and light gray. These new pixels create pseudodetail. See *pixel replication, pseudodetail.*

Pixel Replication A method for increasing the size of a paint file by duplicating existing pixels a set number of times. Necessary to match the size of a file to the number of pixels being printed by a film recorder, and sometimes a printer. See *pixel interpolation.*

Ports Plugs or connectors into which cables are attached.

Program Special computer files that translate instructions given to a computer into a form the computer can understand, and that translates information generated by a computer into a form you, or computer equipment, can understand.

Prompt A symbol that appears on the computer monitor next to the drive specifier to notify you that the computer is waiting for instructions. With many IBMs and compatibles, the prompt is >.

Pseudodetail False detail created by pixel interpolation. Pseudodetail is often impossible to distinguish from actual detail.

Pseudopixels Pixels created by pixel replication or pixel interpolation.

RAM Random Access Memory, a type of chip used to manipulate information.

RAM Disk A method for using RAM as if it were a disk drive. This type of drive is very fast working, but volatile. Once the computer is shut off, all the information in a RAM disk is lost. Also called a *virtual disk* or *RAM drive.*

RAM Drive Same as *RAM disk.*

Random Access The ability to read or write information scattered about the surface of a hard or floppy disk.

Read/Write Head(s) Magnetic devices used to put information on, or get information from, hard or floppy disks.

Removable Mass Storage Device A mass storage device with removable disks. A floppy disk is also removable, but does not store enough information to be considered mass storage.

RGB Abbreviation for Red, Green, Blue, and refers to how many color graphics monitors and many industrial grade TV cameras create images.

RIO Abbreviation for Resolution Independent Object-oriented. For our purposes, a utility program for TIPS.

Sampling The unauthorized use of copyrighted material by using a digital computer.

Saturation Describes whether a color tends toward pastel or being rich. Low-saturation colors are pastel, high-saturated colors are rich in color.

Scanner A device for taking a photographic image and reducing it to numbers that the computer paint program can manipulate.

Screen Resolution The size of the rectangular grid of pixels displayed on a monitor.

Sectors Triangular shaped magnetic dividers on floppy and hard disks, used by the hard/floppy disk controller to rapidly locate different areas of the disk.

Serial Port A plug or connector at the end of a serial transmission device.

Serial Transmission A method by which a computer sends eight bits of information, one bit at a time. When the eight bits are received, the computer interprets them as one byte, and then starts to assemble the next eight bits into a byte. See *parallel transmission.*

Service Bureau The computer graphics equivalent to a photographic lab.

Short Persistence Phosphors The standard phosphorescent coating inside most monitors. See *long persistence phosphors.*

Single Sided (1) A floppy disk that can only have information put on one of its two sides. (2) A floppy disk drive that can only read from or write to one side of a floppy disk.

Software A term for programs. Since programs are really on/off electronic signals, they are considered "soft" or "not physical."

Software Selectable The computer user can make a choice between two or more options by giving a program instructions.

Subsampling If an image is made up of more pixels than the graphics board is capable of sending to the monitor at one time, the entire image can be displayed if some of the pixels that make up the image are left out.

Simultaneously Displayable Colors The total number of colors that the graphics expansion board is capable of sending to the monitor at one time. This may be the same, or less, than the total palette.

Tablet See *digitizing tablet.*

TARGA Abbreviation for Truevision Advanced Raster Graphics Adaptor. The type of graphics expansion board, including compatibles, recommended for use in digital photography.

TIPS Abbreviation for Truevision Image Processing Software. A paint program used for digital photography with TARGA boards.

Total Palette The total number of colors a computer graphics board can make/manipulate. See *simultaneously displayable colors.*

Tracks Circular magnetic area on hard/floppy disks onto which information from the computer can be written to or read from.

TTL (1) Abbreviation for Transistor to Transistor Logic, it is the technology used in monochrome monitors. (2) Abbreviation for Through The Lens light meter.

Upgrade The modification of computers and equipment with outdated technology to work as newer models.

Utility Program A program designed to either enhance or make another program easier to use.

VGA Abbreviation for Video Graphics Array. The graphics standard that replaced EGA.

Video Controller Board An expansion board that operates a monitor.

Video RAM: RAM used for the computer monitor only.

Virtual Drive or Disk Alternate terms for *RAM drive.*

Window For this book, a window is an area marked on the screen which represents a full screen of information. When using large scanned files, the entire image is not viewable on the monitor at one time. By using programs such as

LUMENA or RIO, the entire image can be displayed in a subsampled manner. Some of the pixels that exist on the file will not be displayed on the monitor, making room for the entire image (lacking some of its detail) to be viewed. Most subsampled images will have a window overlaid on it. Any picture area included in the window will fill an entire monitor screen when it is selected to be displayed and worked on.

WORM A removable mass storage device that uses a laser beam to permanently write information to a disk, hence Write Once Read Many.

SOURCES

The information in this section represents products and services useful in digital photography. Keep in mind that prices and product specifications change. Additionally, these listings are not all inclusive, but meant to give you an idea of what products and services are available, and a means to contact companies for further information. All addresses and telephone numbers are in the United States unless otherwise noted.

PUBLICATIONS

Periodicals

Advanced Imaging, PTN Publishing, 210 Crossways Park Drive, Woodbury, New York 11797. Published ten times a year. Subscriptions: free to qualified computer professionals, all others; USA $40 per year; International, add $25 surface, $75 airmail. Articles about video applications with computers. Recent articles and information: digital video cameras for image processing; still video photography; high definition television (HDTV); new products; and a calender of upcoming trade shows of interest to their readers. Technical, but understandable.

AudioVisual Communications, PTN Publishing, 210 Crossways Park Drive, Woodbury, New York 11797. Published monthly. Subscriptions: USA $50 per year; international, $80 per year (international money orders only). Audiovisual orientated. Recent articles and information: slide preparation for video presentations; videotaping at the New York City Metropolitan Opera; digital audio basics; new products. Many of the articles describe how a company uses particular types of equipment. Easy to read.

Byte, McGraw-Hill, Inc., Byte Subscriptions, P.O. Box 551, Hightown, New Jersey 08520. Published 13 times a year; USA $24.95; Canada, $26.95; contact them for rates for other countries. Tel. (800) 257-9402; in New Jersey, (609) 426-5535. Byte is a general purpose computing magazine covering both Macintosh and IBM and compatible computers. The articles cover all aspects of com-puting, and not specifically computer graphics. However, it will keep you abreast of new computing equipment, has review articles of equipment and software, and runs occasional articles that will be of specific interest to digital photographers. While it is not a technical magazine, many of the articles will be difficult to read at first, but as your computer vocabulary increases, so too will the readability of the magazine.

Cadence, Ariel Communications, Inc., 12710 Research, Austin, Texas 78759. Published monthly. Subscriptions: USA $34.95 per year; Canada and Mexico, $44.95; international, $68 (25-day delivery), $108 (10-day delivery). Oriented to computer-assisted design and drafting (CADD) for the IBM and IBM-compatibles and Macintosh computers. Articles are written for professionals, particularly for those who use *AutoCad*, a sophisticated CADD program. This publication will be difficult for beginners to understand.

Computer Graphics World, PennWell Publishing; Circulation Dept., P.O. Box 21638, Tulsa, Oklahoma 74101-9980. Tel. (800) 331-4463 USA only, or (918) 831-9400. $28 USA, $32 Canada/Mexico, $38 international for one year. Computer Graphics World is a general graphics magazine which covers CAD/CAM, video applications in computer graphics, TARGA and TARGA-compatible equipment and information, scanners, etc. Also includes new products and a comprehensive national (USA) and international trade show and conference listing. Although not highly technical and relatively easy to read, new computer users may stumble a bit at first over some of the vocabulary. A good magazine for keeping current on what's happening in computer graphics, and for building your vocabulary.

Computer Pictures Magazine, Montage Publishing, Inc., 25550 Hawthorne Blvd., Suite 314, Torrance, California 90505. Tel. (213) 373-9993. Subscriptions $30 a year. Covers equipment and techniques of interest to digital photographers, including a new products section, "how to" articles, etc. Relatively easy to read.

Digital Photography Newsletter, Suite 5099, 256 S. Robertson Blvd., Beverly Hills, California 90211. Published every four months: three issues, $36 USA and international. Back issues, $6 to subscribers, $12 to non-subscribers. Written specifically for photographers, this newsletter includes information about new products, evaluates products in relation to their use in digital photography and their usefulness to photographers, discusses their strengths and weaknesses, and describes in non-technical terms how they work. The first issue explains in detail some methods for using gray scale scanners to make continuous tone full color TARGA-16 files. Written for the beginning and intermediate level reader. Your best bet to increase your digital photography knowledge and vocabulary.

MicroCAD News, Ariel Communications, Inc., 12710 Research, Austin, Texas 78759. Published monthly. Subscriptions: USA $34.95 per year; Canada and Mexico, $44.95; international, $68 (25-day delivery), $108 (10-day delivery). Tel. orders (512) 250-1700. Articles about CADD applications for IBM and Macintosh computers, including information about desk-top publishing, imaging (which has digital photography applications), new product and upcoming trade show listings. Recent articles and information on: VGA/video input; service bureaus; manufactures of TARGA-type boards and programs; the pros and cons of various operating systems, such as DOS, OS/2, and others. Judging from the questions readers ask to have answered, many of the readers use TARGA-type equipment. Relatively easy to read.

PC, Ziff-Davis Publishing Company. Subscriptions address: PC, P.O. Box 51524, Boulder, Colorado 80321-1524. Tel. (800) 289-0429. Published 22 times a year, USA $44.97; all other countries add $31 per year. PC is a general purpose IBM and IBM-compatible-oriented magazine. It runs review articles of hardware and software products, compares groups of products to each other (for instance, 20 AT-compatible computers under $2,000), and while it is not primarily computer graphics oriented, it does occasionally review and discuss film recorders, graphics scanners, and other equipment that is of interest to digital photographers. As is the case with Byte, it is not very technical, but many of the articles may seem technical to new computer users.

Books

DOS Made Easy ($19.95) by Herbert Schildt; Osburn MacGraw-Hill Publishers. Tel. (800) 227-0900. A good introduction to MS-DOS, and what you can do with it. DOS is the disk operating system for IBM and IBM-compatible computers, discussed briefly in Chapter 2. Although manuals come with the DOS program, they are generally considered not to be of much help to the beginner. The main problem is that they tell you what the different commands do, but do not give enough meaningful examples to help most new users. This book explains DOS, gives examples, and is well organized and easy to read.

IBM AT Clone Buyers Guide and Handbook ($24.95) by Edwin Rutsch; Modular Information Systems, 431 Ashbury St., San Francisco, California 94117. Everything you need to know about AT clones and how to buy them. Easy to read and informative.

Personal Computers Made Easy ($18.95) by James L. Turley; Osburn MacGraw-Hill Publishers. Tel. (800) 227-0900. An excellent beginning book about personal computers, covering all aspects of their care and feeding. Highly recommended for the first time user. Covers IBMs and IBM-compatibles, and Macintosh. Easy to read.

The Best Book of DOS ($24.95) by Alan Simpson; Howard W. Sams & Co. Tel. (800) 428-SAMS. A good book to introduce you to DOS.

The Complete Handbook of Personal Computer Communications ($14.95) by Alfred Glossbrenner, St. Martin's Press, 175 Fifth Avenue, New York, New York 10010. Tel. (800) 221-7945 (in New York, [212]- 674-5151). *Everything* you need to know about modems, and using them to communicate with bulletin board services (BBS). BBSs will allow you to communicate with others who share your interests in digital photography, get public domain (free) utility programs you may find useful, and send and receive digital photographs. Very easy to read and very informative.

The Mac Idiot Book ($18.95) by William W. Flint; Kern International. Tel. (617) 934-2452. A very good first book on Macintosh computers for those of you who wish to do digital photography with a Macintosh. Easy to read.

The Quick and Easy Macintosh Guide for MS-DOS Users ($21.95) by Rusel DeMania and David Altounian; Ashton Tate. Tel. (800) 236-8283. A guide for those who are familiar with DOS and wish to use Macintosh computers (for instance, those of you who decide to use the NuVista board). The book builds on what you already know and shows you the Macintosh equivalent commands.

The Scanner Book ($22.95) by Stephen Beale and James Cavuoto; microPublishing Press, 21150 Hawthorne Blvd. #104, Torrance, California 90503. This book is B&W desktop publishing oriented, but it is extremely easy to read and gives a good overview of the different types of scanners, how they work, and their proper use.

Things the Manual Never Told You ($12.95) by The Boston Computer Society, edited by Jack McGrath; Addison-Wesley Publishing Co. This book acts as a knowledgeable friend, answering often raised questions about personal computers. Full of helpful advice.

MAJOR COMPANIES DISCUSSED IN THIS BOOK

AT&T Graphics Software Labs make RIO and other software. For your nearest authorized reseller, call (317) 844-4364.

Control Systems makes the RealVision-16 and 24 graphics boards discussed in Chapter 4. For information, call them at (612) 631-7800 in the USA, or (44) 0494-26211 in Europe (England).

Time Arts makes LUMENA, ColorScheme I and II, and other software. For your nearest authorized reseller, call (707) 576-7722.

Truevision makes TARGA, Vista and NuVista boards, among other equipment, and sells **TIPS** software. For your nearest authorized reseller, call (317) 841-0332.

Vision Technologies makes Vision graphics boards, sells FreeStyle software, and their parent company (Everex) makes IBM-compatible computers. For your nearest authorized reseller, call (415) 683-2900.

International

Most of the companies listed in this section are USA based. **Techexport Inc.** is an international marketing company which sells many of these products in Europe, including TARGA/Vista, LUMENA, RIO, and Howtek products. USA headquarters: 1 North Avenue, Burlington, Massachusetts 01803, Tel. (617) 229-6900; United Kingdom, Techex Ltd. London, Tel. (44) 1 991 0121; France, Techex SARL, Paris, Tel. (33) 1 3 952-6253; West Germany, Techex GMBH, Munich, Tel. (49) 89 612-0010; Italy, Techex SRL, Milan, Tel. (39) 2 612-8131; Switzerland, Techexport Inc., Zurich, Tel. (41) 1 825-0949.

MAIL ORDER

If you buy a computer by mail, make sure you get a "30 day, no questions asked, money-back guarantee" to assure yourself that the computer does work with the graphics board you choose.

AST is a well-known maker of computers and computer equipment, but does not sell products other than those AST itself makes. Contact them at (714) 863-0181 and ask for a catalog.

CompuAdd Corporation sells IBM/XT/AT clones, hard disks, and various hardware and software products. They have recently begun to open discount stores. This is the only "full" mail order company selling a wide variety of products listed in this section. Contact them at (800) 627-1967.

Everex is a well-known manufacturer of computers and computer equipment, and is the parent company of Vision Technologies, makers of the Vision series of boards. Call Everex at (800) 334-4552 and ask for their catalog.

Northgate sells their computers mail order. Call them at (800) 536-2442 and ask for a catalog.

ZEOS sells their computers at excellent prices, and has recently introduced 24 hour, 365 day telephone assistance, a service that others should, and probably will, emulate. Call at (800) 423-5891 and ask for a catalog .

SERVICE BUREAUS

Bowers Imaging makes murals for use at trade shows and advertising from TARGA-type files. Contact them for more information at 2990 San Pablo Ave., Berkeley, California 94702. Tel. (415) 548-9185.

Datastream. Write or call and ask for a listing of their services: 239 C Southland Dr., Lexington, Kentucky 40503. Tel. (606) 277-5656.

Digital Image will make 35mm slides from TARGA-type files in 2K resolution ($7), or 4 x 5 inch transparencies ($35), or 8 x 10 inch transparencies ($45) at 4K resolution, as well as make 8 x 10 inch Polaroid prints ($22.50 for B&W, $28.50 for color). A 24-hour turnaround time is typical. Contact them at 10526 W. Cermak Rd., Room 304, Westchester, Illinois 60154. Tel. (708) 531-1331.

Electric Image Center. Call or write for a listing of their services: 77 Route 5, Edgewater, New Jersey 07020. Tel. (201) 224-6676.

Imageset Corporation will digitize 35mm slides using a Barneyscan scanner, put TARGA-type files onto 35mm film at 2K resolution ($6.75), and make color separations from a file (from $150 for 4 x 5 inch or smaller). Contact them at 555 19th St., San Francisco, California 94107. Tel. (800) 347-4624.

Jetgraphix specializes in large (up to 30 x 42 inch) ink jet prints, which are made with Fujix Jetgraphy 3000 industrial printers, and also does scanning of prints. Print prices start at about $60, and scans are about $30. They do not do film output. Contact them at 1531 Pontius Ave., Suite 300, Los Angeles, California 90025. Tel. (213) 479-4994.

Kinetic Designs specializes in making IRIS Inkjet prints from TARGA files. For more information, contact them at 64 Fulton St., #1111, New York, New York 10038. Tel. (212) 578-1083.

Speed Graphics makes slides at up to 16K resolution, and scans up to 4K resolution. For more information, contact them at their Computer Services Department, 342 Madison Ave., Suite 411, New York, New York 10173.

Stokes Imaging Services will make 35mm slides from TARGA-type files at 2K resolution ($15) or 4K resolution ($20), as well as make duplicate slides (from 1 to 5000) with fast turnaround times. Contact them at 7000 Cameron Road, P.O. Box 14277, Austin, Texas 78761-4277. Tel. (512) 458-2201.

The Graphic Source will scan TARGA-type images up to 512 pixel resolution ($20), 1K ($30) or 2K ($40) resolutions, make 35mm slides of TARGA-type files at 2K resolution ($14.95), and make color thermal transfer prints ($9.95). Rush service available. Contact them at 350 W. 31st St., 5th Floor, New York, New York 10001. Tel. (212) 268-1990.

Video Technics will make 35mm slides of TARGA-type files at 2K resolution ($35), or 4 x 5 inch transparency ($50) or 8 x 10 inch transparency ($75) at 4K resolution, as well as make color thermal transfer prints ($10). Contact them at 1450 Monroe Drive, P.O. Box 14247, Atlanta, Georgia 30324. Tel. (404) 876-5709.

Western Imaging scans images to TARGA format, makes slides, and performs other services for digital photographers. Write or call and ask for a listing of their services and prices at 1117 Independence Ave., Mountain View, California 94043. Tel. (415) 960-0500.

GRAPHICS BOARDS

Artist RealVision boards, discussed at length in Chapter 4, are manufactured and sold by *Control Systems*. Contact them at 2675 Patton Road, St. Paul, Minnesota 55113. Tel. (612) 631-7800. In Europe (UK), Victoria House, 28/38 Desborough St., High Wycombe, Bucks HP11 2NF England. Tel. (44) 0494-26211.

TARGA and Vista/NuVista boards, discussed at length in Chapter 4, are manufactured by *Truevision*, and sold through authorized representatives. For your nearest reseller, contact Truevision at 7351 Shadeland Station, Suite 100, Indianapolis, Indiana 46256. Tel. (800) 858-TRUE.

Vision boards discussed at length in Chapter 4 are manufactured by *Vision Technologies*, and sold through authorized resellers. For your nearest reseller, contact Vision Technologies at 48504 Kato Road, Fremont, California 94538 Tel. (415) 683-2900.

PAINT AND ANIMATION PROGRAMS

Crystal 3D ($11,995) is a solids modeler and animation package. Crystal 3-D easily integrates with Time Arts LUMENA paint software or with AutoCad files through an optional IGES converter program. Output can be rendered to videotape or film via stop-frame animation controllers. For more information, contact them at 1425 Corporate Center Pkwy., Santa Rosa, California 95407. Tel. (707) 576-7722.

Halovision ($450) from *Media Cybernetics, Inc.* is a paint program that can be used with the TARGA series of graphics boards. Contact them at 8484 Georgia Ave., Silver Spring, Maryland 20910. Tel. (301) 495-3305.

LUMENA (from $2,500 for the TARGA-M8, 16, 32, Vision-16, Realvision-16, and $4,000 for the Vista 2M or 4M) and **ColorScheme II** ($795) for the Vision-16 from *Time Arts* are discussed at length in Chapter 7. LUMENA is for the design and production of professional quality images. Images can be captured from live or recorded video or input via a film or print scanner. Output can be directed to ink jet, laser, and thermal printers or film recorders for high-resolution slides. Contact them at 1425 Corporate Center Pkwy, Santa Rosa, California 95407. Tel. (707) 576-7722.

Parachute ($2,495) from *KB Systems* is a sophisticated animation paint program for use with TARGA-16, 24, and 32. With Parachute, after you paint your image, you can scale, rotate it around a reference point, and move it along a path. You can use the "metamorphosis" function to automatically turn one drawing into another within a specified number of frames. Contact them at 3103 Bee Caves Rd., #235, Austin, Texas 78746. Tel. (512) 328-9366.

RIO ($1,800), discussed in Chapter 7, allows you to draw 2D images and manipulate TARGA-16, 24, and 32 images up to 8K resolution, and control numerous input and output devices. **RIO VISTA** ($1,995) is for the Vista boards. Contact AT&T Graphics Software Labs, 3520 Commerce Crossing, Suite 300, Indianapolis, Indiana 46240. Tel. (317) 844-4364.

SnapShot ($495) from *Aldus Corporation* is a paint program designed for desktop publishing. It contains a number of retouching functions, as well as brightness and contrast controls, three levels of sharpening, posterization effects, and the ability to manipulate 256 gray tones. SnapShot works with both the NTSC TARGA-M8 and Vision-8, but not the PAL versions. Contact them at Aldus Information Center, P.O. Box 25739, Santa Ana, California 92799-9948. Tel. (800) 333-ALDU.

TIPS from *Island Graphics* is sold only through authorized Truevision resellers. For your nearest reseller, contact them at 7351 Shadeland Station, Suite 100, Indianapolis, Indiana 46256. Tel. (800) 858-TRUE.

TOPAS from *AT&T Graphics Software Labs* allows you to create sophisticated 3D drawings and animation. Three modules are available. The *Modeler* ($2,995) works with TARGA-16s, and allows you to draw images, insert up to 20 light sources for highlights and reflections, and output the image at up to 8K resolution; the *ProModeler* ($4,995) works with the TARGA-16, 24, and 32, and includes the features of Modeler but adds features such as metallic (shiny metal) and soft-edged shadows, and works with the TARGA-16, 24, and 32 and Vista boards; *Animator* ($11,995) combines the ProModeler with sophisticated animation capabilities, including automatically generated storyboards, rotoscoping, 3D motion, and path controls. Contact them at 10291 N. Meridian, Suite 275, Indianapolis, Indiana 46290. Tel. (317) 844-4364.

VersaCAD Design ($2,995) from *VersaCAD Corporation* is a computer-aided design program which can be used with the TARGA-16. Contact them at 2124 Main St., Huntington Beach, California 92648. Tel. (714) 960-7720.

UTILITY PROGRAMS

ADI Driver ($450) from *Timchula Creative Limited* is a driver to allow AutoCAD products and video images to be displayed together on a ATVista 4M, for AutoCAD products to be saved in TARGA format and for TARGA format files to be used with AutoCAD products. A video demonstration is available for $20. Contact them at 145 Hudson St., 14th Floor, New York, New York 10013. Tel. (212) 219-1017.

Big D ($995) from *Graphics Software, Inc.* creates photorealistic images from AutoCAD or other CADD programs that create .dfx files. The program creates smooth surface shadings, casts shadows, allows any number of point and directional light sources to be indicated, highlights will be automatically created, and a "light meter" can be used to check the lighting levels at any position of the drawing. Images are saved in TARGA .tga format. Works with TARGA-16 and Vision-16 boards, which means you can incorporate photorealistic computer drawings with digital photographs. Contact them at 9516 Brewster, Dallas, Texas 75772. Tel. (214) 388-8055.

CADVerter ($395) from *AT&T Graphics Software Labs* converts files with the extension .dfx from CAD programs to files that can be used with TOPAS (see Program listings). Contact them at 10291 N. Meridian, Suite 275, Indianapolis, Indiana 46290. Tel. (317) 844-4364.

CYNC Corporation's *ASAvision* ($300) utilities are designed to enhance video input for use with TARGA-16, 24, or 32 boards. Features include Chromakey effects, reduction of electronic noise from video cameras that reduce image quality, and posterization. Contact them at 1482 Beacon St., Brookline, Massachusetts 02146. Tel. (617) 277-4317.

A **driver** for the Shinko CHC-335 and CHC-65 color printers for use with TARGA-type graphics boards is available from *Mitsubishi International*, 520 Madison Ave., New York, New York 10022. Tel. (212) 605-2000.

Graphics Link + ($495) from Harvard Systems Corporation can convert between 17 IBM and Macintosh file types. Contact them at 1661 Lincoln Blvd., Suite 101, Santa Monica, California. Tel. (213) 392-8441.

Image Transformer Plus ($150) from *Micro Visual Systems, Inc.* will convert images from EGA graphics boards to TARGA-16, 24, and 32 formats. Contact them at 385 Elliot St., Newton Upper Falls, Massachusetts 02164. Tel. (617) 964-0640.

Mathematica Print ($95) from *Mathematica, Inc.* is a driver used to make B&W prints from TARGA-type boards on an Hewlett-Packard LaserJet Plus printer. Contact them at P.O. Box 3537, Lakeland, Florida 33802. Tel. (813) 682-1128.

Norman Breslow's Photo Utilities ($65), discussed in Chapter 7, are a group of easy-to-use programs to allow you to lighten and darken images, adjust contrast, make posterizations, change color images to B&W, create color images from B&W, etc. Mail orders only. A sample disk with four working programs, a test image, and complete manual costs $15. Send orders to 12021 Wilshire Blvd., Suite 290, Los Angeles, California 90025.

Pizazz Plus ($149) from *Application Techniques, Inc.* will capture a displayed TARGA-16 or Vision-16 image from your graphics monitor and print it on a Laser Printer or other standard computer printer in up to 256 shades of gray or color (if a color printer is used). The program allows for resizing and cropping of the image before printing. This program also works with a wide variety of EGA/VGA graphics boards. Contact them at 10 Lomar Park Dr., Pepperell, Massachusetts 01463. Tel. (508) 433-5201.

Print TARGA ($335) from *MicroWare, Inc.* are drivers for B&W and color paper printers such as Hewlett-Packard LaserJet, Shinko CHC-335, and Xerox 4020. Program allows for positioning image on the page.

Scan-PCS ($395) also from *MicroWare* is a driver for Sharp/Howtek print scanners for use with TARGA-16, 24, and 32 graphics boards. Contact them at 11329 "P" St., Omaha, Nebraska 68137. Tel. (402) 592-4350.

Ron Scott, Inc. sells a number of utility programs designed for the TARGA series of boards. *QFX* ($500), discussed in Chapter 7, and *QFX-VISTA* ($750) allows you to change color, contrast, and intensity of an image, as well as sharpen, create shadows, and other effects; *SQALE* ($400), mentioned in Chapter 8, uses pixel interpolation to expand files up to 8K resolution or reduce the image to 1/16 of its original size, as well as allow you to convert between differing file formats (e.g., change a TIPS TARGA file to LUMENA's HI RES), and works with TIPS, ATVISTA TIPS, LUMENA, TOPAS, RIO, and Crystal 3D; *QLIC* ($250) prints TARGA and ATVISTA files on a Hewlett-Packard LaserJet printer (and compatibles), allowing you to control the size, placement, contrast and brightness, etc.; *QPR* ($1,500) is a driver for Matrix Procolor, PCR, QCR, and QCR-Z film recorders, automatically expanding the file to 2K, 4K, or 8K resolution. Contact them at 1000 Jackson Boulevard, Houston, Texas 77006. Tel. (713) 529-5868.

ScanRIX (about $700) from *RIX Softworks, Inc.* is a driver for the Sharp JX-450 and JX-300 color scanners, producing files in TARGA-16 and 32 format. Contact them at 18552 MacArthur Blvd., Suite 200, Irvine, California 92715. Tel. (714) 476-8266.

Spectre-Matchprint ($7,500) *Pre-Press Technnologies* makes four color-separation files of TARGA-16, 24, and 32, Vista and NuVista boards which can be used by some computerized offset lithography printers to make color-separation negatives. Contact them at 543 Encinitas Blvd., Suite 114, Encinitas, California 92024. Tel. (619) 753-0194.

TARGA/VISTA Cad Driver ($395) from *AT&T Graphics Software Labs* allows AutoCAD users to merge AutoCAD drawings with images on TARGA-M8, 16, and 24 and Vista boards. Contact them at 10291 N. Meridian, Suite 275, Indianapolis, Indiana 46290. Tel. (317) 844-4364.

TGAPIC ($595) from *Coddbarrett Associates, Inc.* is a driver for Matrix PCR, QCR, and QCR-Z film recorders and TARGA-16, 24, and 32 files. Contact them at 65 Ashburton St., Providence, Rhode Island 02904. Tel. (401) 273-9898.

TIPSDrivers ($445) from *AT&T Graphics Software Labs* are a series of drivers for outputting images from TIPS to film recorders and color thermal and ink jet printers. Contact them at 10291 N. Meridian, Suite 275, Indianapolis, Indiana 46290. Tel. (317) 844-4364.

Western Imaging offers *MAC2TGA* ($299), a conversion program that takes Macintosh I or II files and converts them to TARGA format. Contact them at 1117 Independence Ave., Mountain View, California 94043. Tel. (415) 960-0500.

Videotex Systems, Inc. sells *T-Scan* ($199), a program to convert TARGA-M8 and 16, and Vision 8 and 16 images to files compatible with desktop publishing programs, *Scan-T* ($149) will convert black and white files created by a B&W scanner to TARGA-16 format, and *T-EGA* ($199) will convert video grabbed images from a TARGA or Vision-16 graphics board to EGA format for use with desktop publishing programs. Contact them at 8499 Greenville Ave., Suite 205, Dallas, Texas 75231. Tel. (214) 343-4500.

SCANNERS

Barneyscan ($8,700 IBM) ($9,500 Macintosh) produces TARGA/Vista or NuVista compatible files from 35mm slides or negatives. Expansion board and software to run the Barneyscan are included in the purchase price. Contact them at 1198 10th St., Berkeley, California 94710. Tel. (415) 524-6648.

Datacopy, a Xerox company, makes a number of gray-scale scanners, including the 730GS ($1,300) and the 830 ($2,800). As with most black and white scanners, a conversion utility program is necessary to put the output into TARGA format. For more information, contact them at 1215 Terra Bella Ave., Mountain View, California 94043. Tel. (800) 821-2898 or (415) 965-7900.

Eikonix sells the 1412 scanner (about $27,000) which can digitize images from 16mm film to 40 x 40 inch prints at 4K resolution, and the 1435 ($9,900) 35mm film scanner with up to 4K resolution. Drivers for TARGA-16, 24, and 32 and Vista 2M/4M costs $495. Contact them at 15 Wiggins Ave., Bedford, Massachusetts 01730. Tel. (617) 276-7143.

Howtek sells the Scanmaster, Scanmaster II, Scanmaster 35, all discussed in Chapter 6. Contact them at 21 Park Ave., Hudson, New Hampshire 03051. Tel. (603) 882-5200.

LS-3500 (About $10,000) 35mm film scanner from *Nikon*, discussed in Chapter 6, has a maximum resolution of 6K with a maximum of 16.7M simultaneous colors, and uses either a GPIB or RS-232C serial port. Contact them at Tel. (516) 220-0200, extension 324.

Microtek Lab, Inc. sells a variety of gray-scale flat bed scanners ranging in price from about $2,200 to $4,000 for 256 gray-scale models, as well as software to run them. Different models are available for IBM and IBM clones and Macintosh computers. As is the case with most B&W scanners, a utility program will be necessary to convert the output from the scanner to TARGA format. Microtek has recently announced the MSF-300Z, a color scanner with up to 16.7M simultaneous colors which is compatible with TARGA files. The scanner has not been released for sale at the time of this writing, and no firm price has been given, but it has been reported that the 300Z will sell for about $4,000. For more information, contact them at 680 Knox St., Torrance, California 90502. Tel. (213) 321-2121.

Niscan ($249) from *NISCA Inc.* is a hand scanner with six combinable halftone patterns, three contrast levels and 100, 200, or 300 dots per inch resolution. Software to run the scanner is provided, but conversion software needed to make the file compatible with TARGA-type boards is necessary. Contact them at 1919 Old Denton Rd., Suite 104, Carrollton, Texas 75006. Tel. (800) 245-SCAN.

ScanJet Plus ($2,200) from *Hewlett-Packard* will produce a 256 gray scale with up to 1500 dots per inch resolution. Discussed earlier in this book. Contact them at Hewlett-Packard Business Computing Systems, 19091 Pruneridge Ave., Cupertino, California 95014. Tel. (800) 752-0900.

SC-7500 ($8,950) color scanner from *Mitsubishi*, has a maximum resolution of 400 dpi. For more information contact them at 701 Westchester Ave., White Plains, New York 10604. Tel. (914) 997-4999.

Sharp manufactures the JX-450 and JX-300 flat bed scanners discussed in Chapter 6. They also make the JX-100 (cost about $1,000) hand-held color scanner. Contact them at Sharp Plaza, Mahwah, New Jersey 07430-2135. Tel. (201) 529-8200.

FILM RECORDERS

Celco sells industrial-grade film recorders. The **Professional** model can shoot 16mm to 8 x 10 inch film at 8K resolution, and starts at about $60,000. Contact them at 70 Constantine Drive, Mahwah, New Jersey 07430. Tel. (201) 327-1123. Or, 1150 E. Eighth St., Upland, California 91786. Tel. (714) 985-9868.

LFR (About $10,000) from *Lasergraphics*, discussed in Chapter 8, makes slides up to 4K resolution. IBM, IBM-compatible, and Macintosh models available. The LFR comes with a driver for LUMENA and RIO programs. The *PFR* is a lower cost film recorder (about $6,000) also made by them. Contact them at 17671 Cowan Ave., Irvine, California 92714. Tel. (714) 660-9497.

Matrix (an AGFA company) discussed in Chapter 8, makes numerous film recorders, ranging in price from about $6,000 to $25,000. Contact them at One Ramland Road, Orangeburg, New York 10962. Tel. (800) 852-8533, or in Europe (UK), 0628-25475.

Montage FR1 (about $6,000) film recorder from *Presentation Technologies* produces 4K resolution slides in up to 16.7M colors from TIPS, RIO, LUMENA, ColorScheme I and II, and FreeStyle software, as well as software used with Macintosh computers. Contact them at 743 North Pastoria Ave., Sunnyvale, California 94086. Tel. (800) 345-7050 (in California, 800-749-1959).

Solitaire, from *Management Graphics* discussed in Chapter 8, manufactures film recorders with up to 16K resolution. Contact them at 1401 E. 79th St., Minneapolis, Minnesota 55425. Tel. (612) 854-1220.

I/O

Microsoft Corporation makes the *Microsoft Mouse*, discussed in Chapter 7. They can be bought at almost every computer store, mail order company, or from digital photography resellers.

Summagraphics Corporation makes a number of digitizing tablets suitable for use with TIPS, LUMENA, and similar programs, discussed in Chapter 7. The

Summasketch 961 lists for about $450, and the larger 1201 lists for about $600. Other models are available. For more information, contact them at 60 Silvermine Road, Seymour, Connecticut 06483. Tel. (203) 881-2000.

DECODERS/ENCODERS

Innovative Tech Works sells a decoder ($695) that converts an NTSC video signal to RGB for use with TARGA-type products, and the PC Video Encoder ($895) to change RGB signals to NTSC. Contact them at 405 Battleground Ave., Suite 200, Greensboro, North Carolina 27401. Tel. (919) 370-0855.

Monarch CGC-8708 ($11,900) and **Monarch CGC-8708I** ($14,900) digital scan converter from *Folsom Research, Inc.* takes RGB input information and transforms it to NTSC or PAL output, as well as RGB for viewing on computer video monitors or for use with video projection equipment. Contact them at 526 East Bidwell St., Folsom, Californa 95630. Tel. (916) 985-2481.

VIDI/O Box ($995) from *Truevision* is an encoder/decoder that works with NTSC/RGB and Super VHS and ED Beta formats. Contact AT&T Graphics Software Labs, 10291 N. Meridian, Suite 275, Indianapolis, Indiana 46290. Tel. (317) 844-4364.

CAMERAS

Canon makes a number of still video cameras, including the RC-760 discussed at length in Chapter 5. For more information, contact them at One Canon Plaza, Lake Success, New York 11042-1113. Tel. (800) 892-0020.

JVC Professional Products offer a number of RGB video cameras that can be used with TARGA and Vista-type boards. These cameras have been discussed in Chapter 5. Contact them at 41 Slater Drive, Elmwood Park, New Jersey 07407. Tel. (201) 299-8000.

Pulnix sells a variety of single-chip black and white cameras with resolutions up to 800 TV lines and shutter speeds up to 1/8000 second. Contact them at: (US) 770-A Lucerne Drive, Sunnyvale, California 94086, Tel. (408) 733-1560; (UK) Unit 5, Intec 2, Wade Road, Basingstoke, Hants, RG24 ONE, 0256-475555; (W. Germany) P.O. Box 1202, D-8755 Alzenau, West Germany 06023-4666; (Japan) Takenaka System Co. Ltd. 7-1, Shinomiya Narano-Cho, Yamashina-ku, Kyoto, 607, Tel. 075-593-9787.

Sony makes a number of still video cameras, including the ProMavica discussed at length in Chapter 5. For more information, contact them at Sony Digital Photography/Imaging Dept., Sony Drive, Park Ridge, New Jersey 07656. Tel. (714) 229-4260.

TK 870U ($1,495) is a single-chip RGB 400-TV line video camera, and the **VKC 360** ($1,695) is a single-chip 500 line RGB video camera, both sold by Closed Circuit Products, Tel. (800) 999-3130.

VIDEK MEGAPLUS (from $9,900 to $21,000) from *VIDEK*, a Kodak Company, is a B&W 256 gray-scale digital video camera for use with TARGA-type and Vista/NuVista graphics boards. It has a resolution of 1320 x 1320 pixels for a total of 1.4M pixels. Contact them at 1100 Corporate Drive, Canandaigua, New York 14425-9597. Tel. (800) 24-VIDEK.

PAPER PRINTERS

IRIS Graphics sells the Iris 3024 (from $75,000), a variable dot-size ink jet printer capable of making prints up to 24 x 24 inches, on paper or other (non-stiff) materials. Extremely high quality. Contact them at 12 Jacob Way, Reading, Massachusetts 01867. Tel. (617) 438-1500.

Kodak **XL-7700** (price not announced) is a digital continuous tone printer using thermal dye transfer technology, and that makes full-color prints up to 11 x 11 inches in about 6 minutes. It takes digital information and produces a 2048 x 1536 (35mm format) 8.5 x 11 inch print in full color in about 3.5 minutes, or in black and white in about 90 seconds; and an 11 x 11 inch image has a resolution of 2048 x 2048 pixels. The XL-7700 also prints transparencies, and can increase or reduce (scale up or down) the image file to fit the page size, increase sharpness, etc. Different models for color and B&W, or B&W only, and for IBM, IBM-compatibles and Macintosh computers. *Kodak* **SV-6510** makes continuous tone color prints (up to 16.7M simultaneous colors) or 256 grays from video sources, including those from digital photography graphics boards. Models for IBM-compatibles and Macintosh computers. Ask about prices and drivers/ conversion programs for use with TARGA-type boards. For more information contact Kodak at Tel. (800) 445-6325, ext. 110.

Pixel Master ($7,900) from *Howtek* is a four-color (magenta, yellow, cyan, and black) ink jet printer which uses either a standard RS-232C serial port, or an optional GPIB, and has a resolution of 240 dpi. Contact them at 21 Park Ave., Hudson, New Hampshire 03051. Tel. (603) 882-5200.

QMS Colorgrafix 100 model 65 ($11,995) from *QMS* is a color thermal transfer printer with 300 dot per inch resolution for use with TARGA boards and selected programs. QMS also makes numerous B&W laser printers for IBM, IBM clones and Apple computers. Contact them at P.O. Box 81250, Mobile, Alabama 36689. Tel. (205) 633-4300.

Seiko TARGAPLOT I ($4,995) produces 8 1/2 x 11 inch color thermal prints and the TARGAPLOT II ($5,995) produces 11 x 17 inch color thermal prints from TARGA-type boards. Price includes the driver necessary to run the printer. Numerous other printers are available. Contact them at 1130 Ringwood Court, San Jose, California 95131. Tel. (408) 943-9100 or (408) 922-5800.

Sharp makes color and B&W printers, including the JX-730 ($2,195) color ink jet printer. They can be contacted at (201) 529-8200.

Shinko makes numerous color printers. They can be contacted at Tel. (212) 605-2343 or (213) 977-3727.

Tektronix is a leading manufacturing company involved in computer graphics. They make numerous thermal transfer color printers, ranging in price from moderate to expensive. They can be contacted at the following numbers: USA Tel. (800) 225-5434; France, Africa 33 (169) 07 78 27; West Germany 49 (221) 77220; UK 44 (6284) 6000; Northern Europe 41 (42) 319192; Middle East 43 (222) 68 66 02 0; Canada (705) 737-2700.

Versatec (a Xerox Company) is a leading manufacturing company involved in computer graphics. They make numerous thermal transfer printers. They can be contacted at the following numbers: USA (800) 538-6477 (in California, [800] 341-6060); Canada (416) 494-8204; UK 44-635-517200; France 33-1-6446-1414; West Germany 49-89-61-3000-12; The Netherlands 31-3406-64504; Belgium 32-2-771-9907; Japan 81-3-342-7481.

MISCELLANEOUS

A-VIDD Electronics Company sells a complete line of Sony and Canon still video cameras, recorders and players, video projectors, and are consultants for and operate multimedia presentations for trade shows, sales meetings, etc. Contact them at 3920 Gilman St., Long Beach, California 90815. Tel. (213) 498-7776 or (714) 527-0965.

Mavica Visual Editor ($750) from *GESI/Diaquest* combines still video imaging with electronic presentations and data bases. It allows storing up to 800 still video images on one 2 inch still video floppy disk. Contact them at 1440 San Pablo Ave., Berkeley, California 94702. Tel. (415) 526-7167.

Network Technologies Inc. offers the *VOPEX 2T* ($334) and *8T* ($999), which enables either two or eight monitors to work from the same expansion port. Useful when showing work to clients or giving demonstrations to groups. Contact them at 19145 Elizabeth, Aurora, Ohio 44202. Tel. (800) RGB-TECH.

Send->It (from $1,725) from *Xenas Communications Corporation* is a combination modem plus software program capable of very high transmission speeds, and gives the user the ability to draw on the receiving computer's monitor, among other features. Contact them at 354 Gest St., Cincinnati, Ohio 45203. Tel. (513) 621-2729.

T-Shirt Transfer System ($17,500) from *VisionBase, Inc.* is a complete system for capturing and transferring images to T-shirts and other media. Contact them at 674 Route 202-206 North, Bridgewater, New Jersey 08807. Tel. (201) 218-0900.

Fotovix II-X ($1,800) from Tamron Industires, Inc., is a video camera designed to input slides and negatives to video tape recorders and graphics boards. It was discussed in Chapter 5. For more information, contact them at P.O. Box 388, Port Washington, New York 11050. Tel. (516) 484-8880.

ANIMATION HARDWARE

BCD Associates makes a variety of videotape controllers that allow animated images to be accurately placed onto videotape. The **BCD-4000** (base price $1,750) uses an expansion board and the **BCD-5000** (from $2,495) uses a standard serial port to communicate with the computer. Both of these single frame video controllers use simple commands such as PLAY, STOP, EDIT, etc., and work with a variety of animation programs. Contact them at 7510 N. Broadway, Suite 205, Oklahoma City, Oklahoma 73116. Tel. (405) 843-4574.

DQ-422 ($3,280) from *Diaquest Inc.* is a board that controls a variety of professional NTSC or PAL videotape recorders to allow accurate animation output. Features include remote control of the tape recorder, frame accurate editing, step by step on-screen help, insertion of image by frame numbers, etc. Other models available. **Action Animator Software** ($1,560) used with the DQ-422 allows you to capture sequential video input with a TARGA or Vista board, and allows you to grab explicit frames from videotape, set explicit point for editing, has on-screen help, etc. Contact them at 1440 San Pablo Avenue, Berkeley, California 94702. Tel. (415) 526-7167.

INDEX